Digital Audio and Compact Disc Technology

UStrath

Digital Audio and Compact Disc Technology

Third edition

**Edited by Luc Baert,
Luc Theunissen,
Guido Vergult, Jan Maes
and Jan Arts**
Sony Service Centre (Europe)

Focal Press
An imprint of Butterworth-Heinemann Ltd
Linacre House, Jordan Hill, Oxford OX2 8DP

ℛ A member of the Reed Elsevier plc group

OXFORD LONDON BOSTON
MUNICH NEW DELHI SINGAPORE SYDNEY
TOKYO TORONTO WELLINGTON

First published 1988
Second edition 1992
Third edition 1995

© Sony Service Centre (Europe) NV 1988, 1992, 1995

British Library Cataloguing in Publication Data
Digital Audio and Compact Disc
Technology. – 3Rev.ed
  I. Baert, Luc
  621.38932

  ISBN 0 240 51397 5

Library of Congress Cataloguing in Publication Data
Digital audio and compact disc technology / edited by Luc Baert . . .
  [et al.]. – 3rd ed.
  p.   cm.
  Includes index.
  ISBN 0 240 51397 5
  1. Compact discs. 2. Digital audiotape recorders and recording.
  3. Compact disc players. I. Baert, L. (Luc)
  TK7882.C56D53 1995   94-36853
  621.389'3–dc20   CIP

Printed in Great Britain

Contents

Preface

The past century has witnessed a number of inventions and developments which have made music regularly accessible to more people than ever before. Not the least of these were the inventions of the conventional analog phonograph and the development of broadcast radio. Both have undergone successive changes or improvements, from the 78 rpm disc to the 33⅓ rpm disc, and from the AM system to the FM stereo system. These improvements resulted from demands for better and better quality.

More than 10 years ago, another change took place which now enables us to achieve the highest possible audio fidelity yet – the introduction of digital technology, specifically the compact disc. Research and development efforts, concentrated on consumer products, have made the extra-ordinary advantages of digital audio systems easily accessible at home. Sony is proud to have been one of the forerunners in this field, co-inventor of the compact disc digital audio system and inventor of the MiniDisc, which has led to an entirely new level of quality music.

Sony Service Centre (Europe) NV

A Short History of Audio Technology

Early Years: From Phonograph to Stereo Recording

The evolution of recording and reproduction of audio signals started in 1877, with the invention of the phonograph by T. A. Edison. Since then, research and efforts to improve techniques have been determined by the ultimate aim of recording and reproducing an audio signal faithfully, i.e., without introducing distortion or noise of any form.

With the introduction of the gramophone, a disc phonograph, in 1893 by P. Berliner, the original form of our present record was born. This model could produce a much better sound and could also be reproduced easily.

Around 1925 electric recording was started, but an acoustic method was still mainly used in the sound reproduction system: where the sound was generated by a membrane and a horn, mechanically coupled to the needle in the groove in playback. When recording, the sound picked up was transformed through a horn and membrane into a vibration and coupled directly to a needle which cut the groove onto the disc.

Figure 1 shows Edison's original phonograph, patented in 1877, which consisted of a piece of tin foil wrapped around a rotating cylinder.

Vibration of his voice spoken into a recording horn (as shown) caused the stylus to cut grooves into a tin foil. The first sound recording made was Edison reciting 'Mary Had a Little Lamb' (Edison National History Site).

Figure 2 shows the Berliner gramophone, manufactured by US Gramophone Company, Washington, DC. It was hand-powered and required an operator to crank the handle up to a speed of 70 revolutions per minute (rpm) to get a satisfactory playback (Smithsonian Institution).

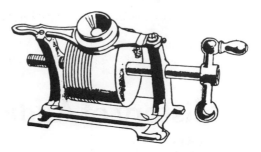

Figure 1 *Edison's phonograph*

Figure 2 *Berliner gramophone*

Further developments such as the electric crystal pick-up and, in the 1930s, broadcast AM radio stations made the SP (standard playing 78 rpm record) popular. Popularity increased with the development, in 1948 by CBS, of the 33⅓ rpm long-playing record (LP), with about 25 minutes of playing time on each side. Shortly after this, the EP (extended play) 45 rpm record was introduced by RCA with an improvement in record sound quality. At the same time, the lightweight pick-up cartridge, with only a few grams of stylus pressure, was developed by companies like General Electric and Pickering.

The true start of progress towards the ultimate aim of faithful recording and reproduction of audio signals was the introduction of stereo records in 1956. This began a race between manufacturers to produce a stereo reproduction tape recorder, originally for industrial master use. However, the race led to a simplification of techniques which, in turn, led to development of equipment for domestic use.

Broadcast radio began its move from AM to FM, with consequent improvement of sound quality, and in the early 1960s stereo FM broadcasting became a reality. In the same period the compact cassette recorder which would eventually conquer the world was developed by Philips.

Developments in Analog Reproduction Techniques

The three basic media available in the early 1960s: tape, record and FM broadcast, were all analog media. Developments since then include:

Developments in turntables

There has been remarkable progress since the stereo record appeared. Cartridges, which operate with stylus pressure of as little as 1 gram were developed and tonearms which could trace the sound groove perfectly with this one gram pressure were also made. The hysteresis synchronous motor and DC servo motor were developed for quieter, regular rotation and elimination of rumble. High-quality heavyweight model turntables, various turntable platters, and insulators were developed to prevent unwanted vibrations from reaching the stylus. With the introduction of electronic technology, full automation was performed. The direct drive system with the electronically controlled servo motor, the BSL motor (brushless and slotless linear motor) and the quartz locked DC servo motor were finally adopted together with the linear tracking arm and electronically controlled tonearms (biotracer). So, enormous progress was achieved since the beginning of the gramophone: in the acoustic recording period, disc capacity was 2 minutes on each side at 78 rpm, and the frequency range was 200 Hz–3 kHz with a dynamic range of

Photo 1 *PS-X75 analog record player*

18 dB. At its latest stage of development, the LP record frequency range is 30 Hz–15 kHz, with a dynamic range of 65 dB in stereo.

Developments in tape recorders

In the 1960s and 1970s, the open reel tape recorder was the instrument used both for record production and for broadcast so efforts were constantly made to improve the performance and quality of the signal. Particular attention was paid to the recording and reproduction heads, recording tape as well as

Photo 2 *TC-766-2 analog domestic reel-to-reel tape recorder*

tape path drive mechanism with, ultimately, a wow and flutter of only 0.02% wrms at 38 cm/s, and of 0.04% wrms at 19 cm/s. Also the introduction of compression/expansion systems such as Dolby, dBx, etc. improved the available signal-to-noise ratios.

Professional open reel tape recorders were too bulky and too expensive for general consumer use, however, but since its invention in 1963 the compact cassette recorder began to make it possible for millions of people to enjoy recording and playing back music with reasonable tone quality and easy operation. The impact of the compact cassette was enormous and tape recorders for recording and playing back these cassettes became quite indispensable for music lovers, and for those who use the cassette recorders for a myriad of purposes such as taking notes for study, recording speeches, dictation, for 'talking letters' and for hundreds of other applications.

Inevitably, the same improvements used in open reel tape recorders eventually found their way into compact cassette recorders.

Limitations of Analog Audio Recording

Despite the spectacular evolution of techniques and the improvements in equipment, by the end of the 1970s the industry had almost reached the level above which few further improvements could be performed without increasing dramatically the price of the equipment. This was because quality, dynamic range, distortion (in its broadest sense) are all determined by the characteristics of the medium used (record, tape, broadcast) and by the processing equipment. Analog reproduction techniques had just about reached the limits of their characteristics.

Figure 3 represents a standard analog audio chain, from recording to reproduction, showing dynamic ranges in the three media: tape, record, broadcast.

Lower limit of dynamic range is determined by system noise and especially the lower frequency component of the noise. Distortion by system non-linearity generally sets the upper limit of dynamic range.

The strength and extent of a pick-up signal from a microphone is determined by the combination of the microphone sensitivity and the quality of the microphone pre-amplifier, but it is possible to maintain a dynamic range in excess of 90 dB by setting the levels carefully. However, the major problems in microphone sound pick-up are the various types of distortion inherent in the recording studio, which cause a narrowing of the dynamic range, i.e., there is a general minimum noise level in the studio, created by, say, artists or technical staff moving around, or the noise due to air currents and breath, and all types of electrically induced distortions.

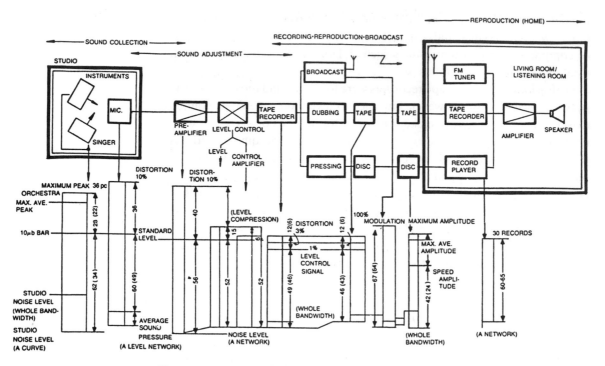

Figure 3 *Typical analog audio systems, showing dynamic range*

Up to the pre-mixing and level amplifiers no big problems are encountered. However, depending on the equipment used for level adjustment, the low *and* high limits of dynamic range are affected by the use of equalization. The type and extent of equalization depend on the medium. Whatever, control amplification and level compression are necessary, and this affects the sound quality and the audio chain. Furthermore, if you consider the fact that for each of the three media (tape, disc, broadcast) master tape and mother tape are used, you can easily understand that the narrow dynamic range available from conventional tape recorders becomes a 'bottle neck' which affects the whole process.

To summarize, in spite of all the spectacular improvements in analog technology, it is clear that the original dynamic range is still seriously affected in the analog reproduction chain.

Similar limits to other factors affecting the system: frequency response, signal-to-noise ratio, distortion, etc. exist simply due to the analog processes involved. These reasons prompted manufacturers to turn to digital techniques for audio reproduction.

First Development of PCM Recording Systems

The first public demonstration of pulse code modulated (PCM) digital audio was in May 1967, by NHK (Japan Broadcasting Corporation) and the record medium used was a 1-inch, 2-head, helical scan VTR. The impression gained by most people who heard it was that the fidelity of the sound produced by the digital equipment could not be matched by any conventional tape recorder. This was mainly because the limits introduced by the conventional tape recorder simply no longer occurred.

As shown in Figure 4a, the main reason why conventional analog tape recorders cause such a deterioration of the original signal is firstly that the magnetic material on the tape actually contains distortion components before anything is actually recorded. Secondly, the medium itself is non-linear, that is, it is not capable of recording and reproducing a signal with total accuracy. Distortion is, therefore, built-in to the very heart of every analog tape recorder. In PCM-recording (Figure 4a), however, the original bit value pattern corresponding to the audio signal, and thus the audio signal itself, can be fully recovered, even if the recorded signal is distorted by tape non-linearities and other causes.

After this demonstration at least, there were no grounds for doubting the high sound quality achievable by PCM techniques. The engineers and music lovers who were present at this first public PCM playback demonstration, however, had no idea when this equipment would be commercially available,

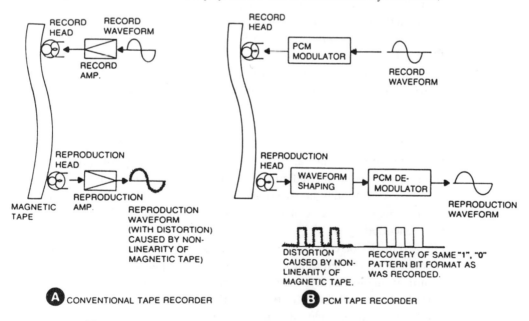

Figure 4 *Showing (a) conventional analog and (b) PCM digital tape recording*

and many of these people had only the vaguest concept of the effect which PCM recording systems would have on the audio industry. In fact, it would be no exaggeration to say that owing to the difficulty of editing, the weight, size, price, and difficulty in operation, not to mention the necessity of using the highest-quality ancillary equipment (another source of high costs), it was at that time very difficult to imagine that any meaningful progress could be made.

Nevertheless, highest-quality record production at that time was by the direct cutting method, in which the lacquer master is cut without using master and mother tapes in the production process: the live source signal is fed directly to the disc cutting head after being mixed. Limitations due to analog tape recorders were thus side-stepped. Although direct cutting sounds quite simple in principle, it is actually extremely difficult in practice. First of all, all the required musical and technical personnel, the performers, the mixing and cutting engineers, have to be assembled together in the same place at the same time. Then the whole piece to be recorded must be performed right through from beginning to end with no mistakes, because the live source is fed directly to the cutting head.

If PCM equipment could be perfected, high-quality records could be produced while solving the problems of time and value posed by direct cutting. PCM recording meant that the process after the making of the master tape could be completed at leisure.

In 1969, Nippon Columbia developed a prototype PCM recorder, loosely based on the PCM equipment originally created by NHK: a 4-head VTR with 2-inch tape was used as a recording medium, with a sampling rate of 47.25 kHz using a 13-bit linear analog-to-digital converter. This machine was the starting-point for the PCM recording systems which are at present marketed by Sony, after much development and adaptation.

Development of Commercial PCM Processors

In a PCM recorder, there are three main parts: an encoder which converts the audio source signal into a digital PCM signal, a decoder to convert the PCM signal back into an audio signal and, of course, there has to be a recording medium, using some kind of magnetic tape for record and reproduction of the PCM encoded signal.

The time period occupied by one bit in the stream of bits composing a PCM encoded signal is determined by the sampling frequency and the number of quantization bits. If, say, a sampling frequency of 50 kHz is chosen (sampling period 20 μs), and that a 16-bit quantization system is used, then the time period occupied by one bit when making a two-channel recording will be about 0.6 μs. In order to ensure the success of the recording, detection bits for the error-correction system will also have to be included. As a result, it is necessary to employ a record/reproduction system which has a bandwidth of between about 1 and 2 MHz.

Bearing in mind this bandwidth requirement, the most suitable practical recorder is a video tape recorder (VTR). The VTR was specifically designed for recording TV pictures, in the form of video signals. To successfully record a video signal, a bandwidth of several megahertz is necessary, and it is a happy coincidence that this makes the VTR eminently suitable for recording a PCM encoded audio signal.

The suitability of the VTR as an existing recording medium meant that the first PCM tape recorders were developed as two-unit systems comprising a VTR and a digital audio processor. The latter was connected directly to an analog hi-fi system for actual reproduction. Such a device, the PCM-1, was first marketed by Sony in 1977.

In the following year, the PCM-1600 digital audio processor for professional applications was marketed. In April 1978, the use of 44.056 kHz as a sampling frequency (the one used in the above-mentioned models) was accepted by the AES (Audio Engineering Society).

At the 1978 CES (Consumer Electronics Show) held in the USA, an unusual display was mounted. The most famous names among the American speaker manufacturers demonstrated their latest products using a PCM-1 digital audio processor and a consumer VTR as the sound source. Compared

Photo 3 *PCM-1 digital audio processor*

with the situation only a few years ago, when the sound quality available from tape recorders was regarded as being of relatively low standard, the testing of speakers using a PCM tape recorder marked a total reversal of thought. The audio industry had made a major step towards true fidelity to the original sound source, through the total redevelopment of the recording medium which used to cause most degradation of the original signal.

At the same time a committee for the standardization of matters relating to PCM audio processors using consumer VTRs was established, in Japan, by 12 major electronics companies. In May 1978 they reached agreement on the EIAJ (Electronics Industry Association of Japan) standard. This standard basically agreed on a 14-bit linear data format for consumer digital audio applications.

The first commercial processor for domestic use according to this EIAJ standard, which gained great popularity, was the now famous PCM-F1 launched in 1982. This unit could be switched from 14-bit into 16-bit linear coding/decoding format so, in spite of being basically a product designed for the demanding hi-fi enthusiast, its qualities were so outstanding that it was immediately used on a great scale in the professional audio recording business as well, thus quickening the acceptance of digital audio in the recording studios.

In the professional field the successor to the PCM-1600, the PCM-1610, used a more elaborate recording format than EIAJ and consequently necessitated professional VTRs based on the U-Matic standard. It quickly became a de facto standard for two-channel digital audio production and compact disc mastering.

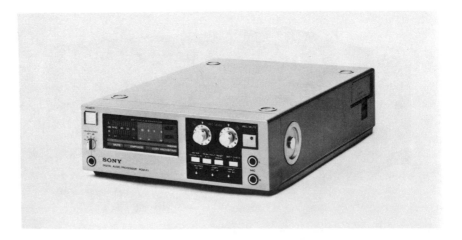

Photo 4 *PCM-F1 digital audio processor*

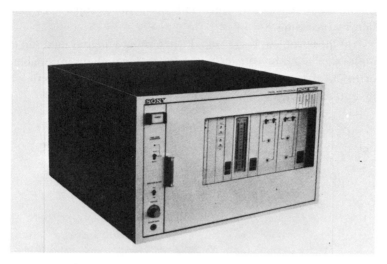

Photo 5 *PCM-1610 digital audio processor*

Stationary Head Digital Tape Recorders

The most important piece of equipment in the recording studio is the multi-channel tape recorder: different performers are recorded on different channels – often at different times – so that the studio engineer can create the required 'mix' of sound before editing and dubbing. The smallest number of channels used is generally 4, the largest 32.

A digital tape recorder would be ideal for studio use because dubbing (re-recording of the same piece) can be carried out more or less indefinitely. On an analog tape recorder (Figure 5), however, distortion increases with each dub. Also, a digital tape recorder is immune to cross-talk between channels, which can cause problems on an analog tape recorder.

It would, however, be very difficult to satisfy studio standard requirements using a digital audio processor combined with a VTR. For a studio, a fixed head digital tape recorder would be the answer. Nevertheless, the construction of a stationary head digital tape recorder poses a number of special problems. The most important of these concerns the type of magnetic tape and the heads used.

The head-to-tape speed of a helical scan VTR (Figure 5) used with a digital audio processor is very high, around 10 metres per second. However, on a stationary head recorder, the maximum speed possible is around 50 centimetres per second, meaning that information has to be packed much more closely on the tape when using a stationary head recorder; in other words, it has to be capable of much higher recording densities. As a result of this, a great deal of research was carried out in the 1970s into new types of modulation recording systems, and special heads capable of handling high-density recording.

Another problem is generated when using a digital tape recorder to edit audio signals – it is virtually impossible to edit without introducing 'artificial' errors in the final result. Extremely powerful error-correcting codes were invented capable of eliminating these errors.

Photo 6 *PCM-3324 digital audio stationary head (DASH) recorder*

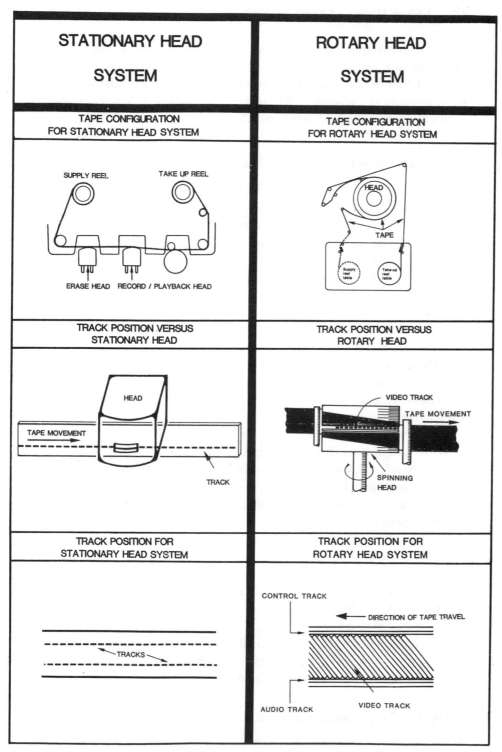

Figure 5 *Analog audio and video tape recording*

The digital multi-channel recorder had finally developed after all the problems outlined above had been resolved. A standard format for stationary head recorders, called DASH (digital audio stationary head) was agreed upon by major manufacturers like Studer, Sony and Matsushita. Example of such a machine is the 24-channel Sony PCM-3324.

Development of the Compact Disc

In the 1970s the age of the video disc began, with three different systems being pursued: the optical system, where the video signal is laid down as a series of fine grooves on a sort of record, and is read off by a laser beam; the capacitance system, which uses changes in electrostatic capacitance to plot the video signal; and the electrical system, which uses a transducer. Engineers then began to think that because the bandwidth needed to record a video signal on a video disc was more than the one needed to record a digitized sound signal, similar systems could be used for PCM/VTR recorded material. Thus the digital audio disc (DAD) was developed, using the same technologies as the optical video discs: in September 1977, Mitsubishi, Sony and Hitachi demonstrated their DAD systems at the Audio Fair. Because everyone knew that the new disc systems would eventually become widely used by the consumer, it was absolutely vital to reach some kind of agreement on standardization.

Furthermore, Philips from the Netherlands, who had been involved in the development of video disc technology since the early 1970s had by 1978 also developed a DAD, with a diameter of only 11.5 cm, whereas most Japanese manufacturers were thinking of a 30 cm DAD, in analogy with the analog LP. Such a large record however would hold up to 15 hours of music, so it would be rather impractical and expensive.

During a visit of Philips executives to Tokyo, Sony was confronted with the Philips idea, and they soon joined forces to develop what was to become the now famous compact disc, which was finally adopted as a world-wide standard. The eventual disc size was decided upon as 12 cm, in order to give it a capacity of 74 minutes: the approximate duration of Beethoven's Ninth Symphony.

The compact disc was finally launched on the consumer market in October 1982, and in a few years, it gained great popularity with the general public, becoming an important part of the audio business.

Peripheral Equipment for Digital Audio Production

It is possible to make a recording with a sound quality extremely close to the original source, when using a PCM tape recorder, as digital tape recorders do not 'colour' the recording, a failing inherent in analog tape recorders. More important, a digital tape recorder offers scope for much greater freedom and flexibility during the editing process.

There follows a brief explanation of some peripheral equipment used in a studio or a broadcasting station as part of a digital system for the production of software.

● **Digital mixer.** A digital mixer which processed the signal fed to it digitally would prevent any deterioration in sound quality, and would allow the greatest freedom for the production of software. The design and construction of a digital mixer is an extremely demanding task. However, multi-channel mixers suitable for use in studios and broadcasting stations have been produced and are starting to replace analog mixing tables in demanding applications.

● **Digital editing console.** One of the major problems associated with using a VTR-based recording system is the difficulty in editing. The signal is recorded onto a VTR cassette, which means that normal cutting and splicing of the tape for editing purposes is impossible. Therefore, one of the most pressing problems after development of PCM recording systems is the design of an electronic editing console. The most popular editing console associated with the PCM-1610 recording system is Sony's DAE-1100.

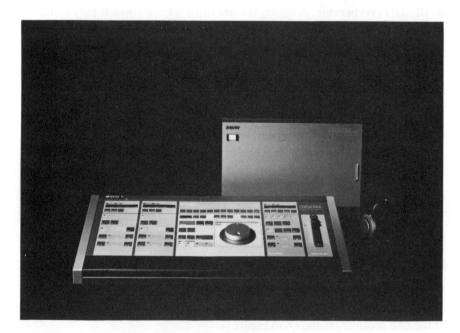

Photo 7 *Digital editing console*

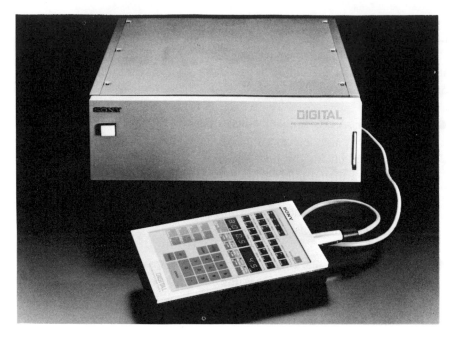

Photo 8 *Digital reverberator*

● **Digital reverberator.** A digital reverberator is based on a totally different concept from conventional reverb units, which mostly use a spring or a steel plate to achieve the desired effect. Such mechanical reverbs are limited in the reverb effect and suffer significant signal degradation. Reverb effect available from a digital reverberation unit covers an extremely wide and precisely variable range, without signal degradation.

● **Sampling frequency conversion and quantization processing.** A sampling frequency unit is required to connect together two pieces of digital recording equipment which use different sampling frequencies. Similarly, a quantization processor is used between two pieces of equipment using different quantization bit numbers. These two devices allow free transfer of information between digital audio equipment of different standards.

Outline of a Digital Audio Production System

Several units from the conventional analog audio record production system can already be replaced by digital equipment in order to improve the quality of the end product, as shown in Figure 6.

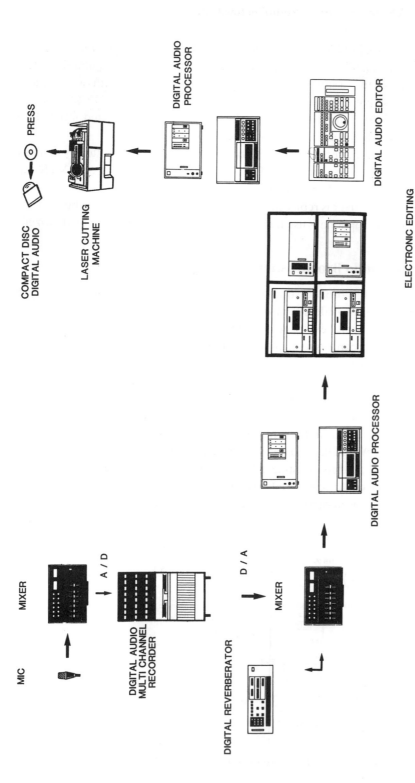

Figure 6 *Digital audio editing and record production*

Audio signals from the microphone, after mixing, are recorded on a multi-channel digital audio recorder. The output from the digital recorder is then mixed into a stereo signal through the analog mixer, with or without the use of a digital reverberator.

The analog output signal from the mixer is then converted into a PCM signal by a digital audio processor and recorded on a VTR.

Editing of the recording is performed on a digital audio editor by means of digital audio processors and VTRs. The final result is stored on a VTR-tape or cassette. The cutting machine used is a digital version.

When the mixer is replaced by a digital version and – in the distant future – a digital microphone is used, the whole production system will be digitized.

Digital Audio Broadcasting

Since 1978, FM broadcasting stations have expressed a great deal of interest in digital tape recorders, realizing the benefits they could bring almost as soon as they had been developed. Figure 7 shows an FM broadcast set-up using digital tape recorders to maintain high-quality broadcasts.

In the near future, high-quality broadcasts will probably be made through completely digitized PCM systems, as shown in Figure 8. A bandwidth of several megahertz would be needed, however, a much higher frequency range than used at present in FM broadcasting. To broadcast such a wide bandwidth, a transmitting frequency with a wavelength of less than one centimetre would have to be used. Because of these factors, the most effective method for broadcasting a PCM signal, also from the point of view of areas which could be covered, would probably be via satellite. A great deal of research has been carried out in this field in recent years.

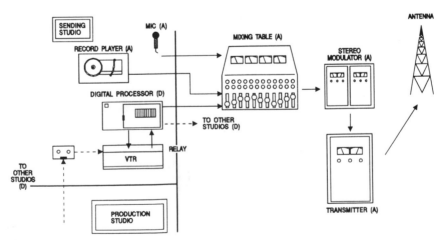

Figure 7 *Mixed analog and digital audio broadcasting system*

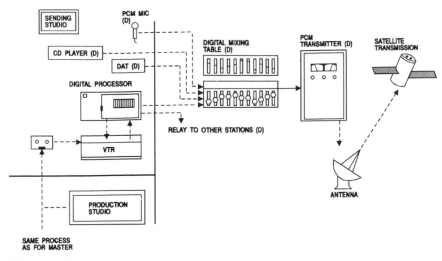

Figure 8 *Digital audio broadcasting system*

A standard for multi-channel digital audio broadcasting has been developed jointly by manufacturers and broadcasters from various countries.

R-DAT and S-DAT: Digital Audio Tape Recorder Formats

Further investigation to develop small, dedicated, digital audio recorders which would not necessitate a video recorder has led to a parallel development of both rotary-head and stationary-head approaches, resulting in the so-called R-DAT (rotary head digital audio tape recorder), and S-DAT (stationary head digital audio tape recorder) formats.

Like its professional counterpart, the S-DAT system relies on multiple-track thin-film heads to achieve a very high packing density, whereas the R-DAT system is actually a miniaturized rotary-head recording system, similar to video recording systems, optimized for audio applications.

The R-DAT system, launched first on the market, uses a small cassette of only $73 \times 54 \times 10.5$ mm – about one half of the well-known analog compact cassette. Tape width is 3.81 mm – about the same as the compact cassette. Other basic characteristics of R-DAT are:

● multiple functions: a variety of quantization modes and sampling rates; several operating modes such as extra-long playing time, 2- or 4-channel recording, direct recording of digital broadcasts
● very high sound quality can be achieved using a 48 kHz sampling frequency and 16-bit linear quantization
● high-speed search facilities
● very high recording density, hence reduced running cost (linear tape speed only 0.815 cm/s)

● very small mechanism
● SCMS: Serial Copy Management System, a copy system that allows the user to make a single copy of a copyrighted digital source.

A description of the R-DAT format and system is given in Chapter 17.

The basic specifications of the S-DAT system have been initially determined: in this case, the cassette will be reversible (as the analog compact cassette), using the same 3.81-mm tape.

Two or four audio channels will be recorded on 20 audio tracks. The width of the tracks will be only 65 μm (they are all recorded on a total width of only 1.8 mm!) and, logically, various technological difficulties hold up its practical realization.

MiniDisc and DCC: New Digital Audio Recording Media

R-DAT and S-DAT have never established themselves in the mass consumer market. Expensive equipment and lack of pre-recorded tapes have led consumers to stay with the analog compact cassette, although broadcasters and professional users favour the R-DAT format because of its high sound quality. This situation has led to the development of two new consumer digital audio formats, designed to be direct replacements of the analog compact cassette. The requirements for the new formats are high. The recording media must be small and portable with a sound quality approaching the sound quality of a CD. The formats must receive strong support from the software industry for pre-recorded material. Above all, the equipment must be cheap.

The answer from Sony is the MiniDisc, a 6.4 cm disc housed in a rigid protective cartridge. A disc-based system has the advantage of quick access to any part of the music, a feature that all CD users appreciate. The optical read-out system with non-contact recording and playback capability make the MiniDisc an extremely durable and long-lasting sound medium. The 4-Mbit semiconductor memory enables the listener to enjoy uninterrupted sound, even when the set vibrates heavily. This 'shockproof' memory underlines the portable nature of MiniDisc. The MiniDisc format is described in detail in Chapter 19.

The reply from Philips is the digital compact cassette (DCC). As the name suggests, DCC has a lot in common with the analog compact cassette which was introduced almost 30 years ago by the same company. The DCC cassette basically has the same dimensions as the analog compact cassette, which enables downward compatibility: new DCC machines can still playback analog compact cassettes. The DCC system uses a stationary thin film head with 18 parallel tracks, comparable with the now obsolete S-DAT format. Tape width and tape speed are equal to the analog compact cassette. A description of the DCC format can be found in Chapter 20.

Although DCC and MiniDisc represent very different approaches to digital recording formats for the consumer market, they also have something in common. For the first time in digital audio signal processing special coding techniques are used to reduce the source bit rate. MiniDisc uses the so-called adaptive transform acoustic coding (ATRAC) data compression to achieve a data reduction of about one fifth of the original amount of data. DCC uses precision adaptive sub-band coding (PASC) to reduce the amount of data to about one fourth of the source data. Both compression techniques make use of psycho-acoustic principles of the human ear. The human ear can only hear sound above a certain level, this threshold level depends on the frequency of the sound. Another effect is simultaneous masking where a loud sound can render a soft sound inaudible, e.g. like a conversation that becomes inaudible when a train passes by. Data reduction is achieved by removing inaudible sounds from the signal, taking into account these psycho-acoustic principles.

Summary of Development of Digital Audio Equipment at Sony

October 1974

First stationary digital audio recorder, the X-12DTC, 12-bit

September 1976

FM format 15-inch digital audio disc read by laser. Playing time: 30 minutes at 1800 rpm, 12-bit, 2-channel

October 1976

First digital audio processor, 12-bit, 2-channel, designed to be used in conjunction with a VTR

June 1977

Professional digital audio processor PAU-1602, 16-bit, 2-channel, purchased by NHK (Japan Broadcasting Corporation)

September 1977

World's first consumer digital audio processor PCM-1, 13-bit, 2-channel 15-inch digital audio disc read by laser. Playing time: 1 hour at 900 rpm

March 1978

Professional digital audio processor PCM-1600

Photo 9 *DTC-55ES R-DAT player*

April 1978

Stationary-head digital audio recorder X-22, 12-bit, 2-channel, using ¼-inch tape

World's first digital audio network

October 1978

Long-play digital audio disc read by laser. Playing time: 2½ hours at 450 rpm

Professional, multi-channel, stationary-head audio recorder PCM-3224, 16-bit, 24-channel, using 1-inch tape

Professional digital audio mixer DMX-800, 8-channel input, 2-channel output, 16-bit

Professional digital reverberator DRX-1000, 16-bit

May 1979

Professional digital audio processor PCM-100 and consumer digital audio processor PCM-10 designed to EIAJ (Electronics Industry Association of Japan) standard

October 1979

Professional, stationary-head, multi-channel digital audio recorder PCM-3324, 16-bit, 24-channel, using ½-inch tape

Professional stationary-head digital audio recorder PCM-3204, 16-bit, 4-channel, using ¼-inch tape

May 1980

Willi Studer of Switzerland agrees to conform to Sony's digital audio format on stationary-head recorder

June 1980

Compact disc digital audio system mutually developed by Sony and Philips

October 1980

Compact disc digital audio demonstration with Philips, at Japan Audio Fair
in Tokyo

February 1981

Digital audio mastering system including digital audio processor PCM-1610,
digital audio editor DAE-1100 and digital reverberator DRE-2000

Spring 1982

PCM adapter PCM-F1 which makes digital recordings and playbacks on a
home VTR

October 1982

Compact disc player CDP-101 is launched onto the Japanese market and as of
March 1983 it is available in Europe

1983

Several new models of CD players with sophisticated features: CDP-701,
CDP-501, CDP-11S
PCM-701 encoder

November 1984

Portable CD player: the D-50
Car CD players: CD-X5 and CD-XR7

1985

Video 8 multi-track PCM-recorder (EV-S700)

March 1987

First R-DAT player DTC-1000ES launched onto the Japanese market

July 1990

Second generation R-DAT player DTC-55ES available on the European
market

March 1991

DAT Walkman: TCD-D3
Car DAT player: DTX-10

November 1992

World-wide introduction of the MiniDisc (MD) format, first generation
MD Walkman: MZ-1 (recording and playback) and MZ-2P (playback
only)

PART ONE

Principles of Digital Signal Processing

1
Introduction

For many years, two main advantages of the digital processing of analog signals have been known.

First, if the transmission system is properly specified and dimensioned, transmission quality is independent of the transmission channel or medium. This means that, in theory, factors which affect the transmission quality (noise, non-linearity, etc.) can be made arbitrarily low by proper dimensioning of the system.

Second, copies made from an original recording in the digital domain, are identical to that original; in other words, a virtually unlimited number of copies, which all have the same basic quality as the original, can be made. This is a feature totally unavailable with analog recording.

A basic block diagram of a digital signal processing system is shown in Figure 1.1.

Digital audio processing does require an added circuit complexity, and a larger bandwidth than analog audio processing systems, but these are minor disadvantages when the extra quality is considered.

Perhaps the most critical stages in digital audio processing are the conversions from analog to digital signals, and vice versa. Although the principles of A/D and D/A conversion may seem relatively simple, in fact, they are technically speaking very difficult and may cause severe degradation of the original signal. Consequently, these stages often generate a limiting factor that determines the overall system performance.

Conversion from analog to digital signals is done in several steps:

- filtering – this limits the analog signal bandwidth, for reasons outlined below
- sampling – converts a continuous-time signal into a discrete-time signal

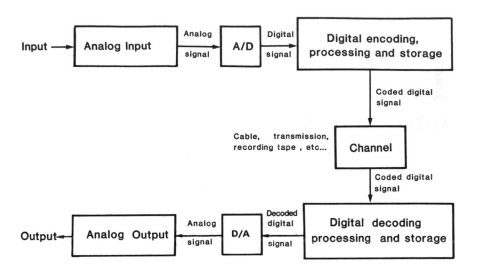

Figure 1.1 *Basic digital signal processing system*

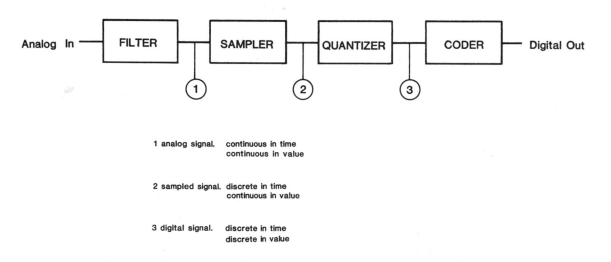

Figure 1.2 *An analog-to-digital conversion system*

● quantization – converts a continuous-value signal into a discrete-value signal

● coding – defines the code of the digital signal according to the application that follows.

Figure 1.2 shows the block diagram of an analog-to-digital conversion system.

2
Principles of Sampling

The Nyquist Theorem

By definition, an analog signal varies continuously with time. To enable it to be converted into a digital signal, it is necessary that the signal is first sampled, i.e., at certain points in time a sample of the input value must be taken (Figure 2.1). The fixed time intervals between each sample are called **sampling intervals** (t_s).

Although the sampling operation may seem to introduce a rather drastic modification of the input signal (as it ignores all the signal changes that occur between the sampling times), it can be shown that the sampling process in principle removes *no information whatsoever*, as long as the sampling frequency is at least present in the input signal. This is the famous **Nyquist theorem** on sampling (also called the Shannon theorem).

The Nyquist theorem can be verified if we consider the frequency spectra of the input and output signals (Figure 2.2).

An analog signal $i_{(t)}$ which has a maximum frequency f_{max}, will have a

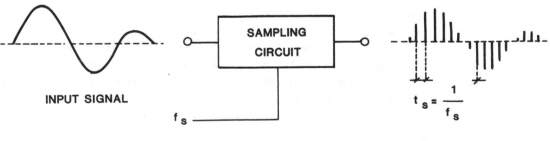

Figure 2.1

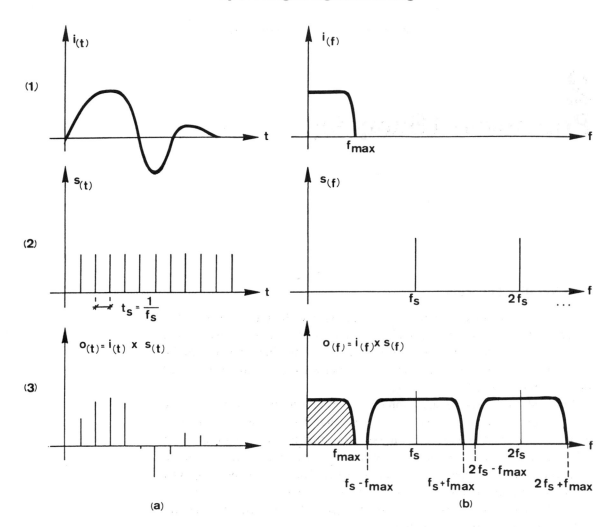

Figure 2.2 *Showing the sampling process in (a) time domain and (b) frequency domain*

spectrum having any form between 0 Hz and f_{max} (Figure 2.2.1a); the sampling signal $s_{(t)}$, having a fixed frequency f_s, can be represented by one single line at f_s (Figure 2.2.1b). The sampling process is equivalent to a multiplication of $i_{(t)}$ and $s_{(t)}$, and the spectrum of the resultant signal (Figure 2.2.3b) can be seen to contain the same spectrum as the analog signal, together with repetitions of the spectrum modulated around multiples of the sampling frequency. As a consequence, low-pass filtering can completely isolate and thus completely recover the analog signal.

The pictures in Figure 2.3 show two sine waves (bottom traces) and their sampled equivalents (top traces).

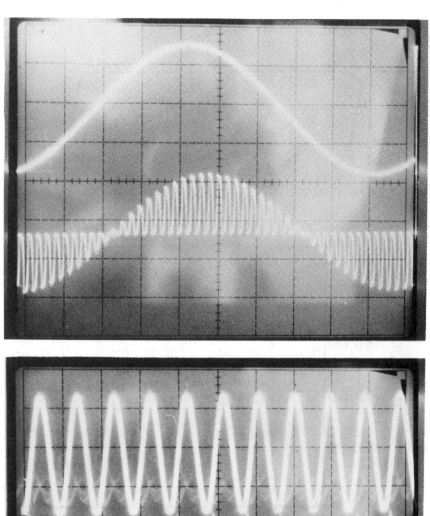

Figure 2.3 *Two examples of sine waves together with sampled versions (a)* (top) *1 kHz sine wave (b)* (bottom) *10 kHz sine wave. Sampling frequency f_s, is 44.056 kHz*

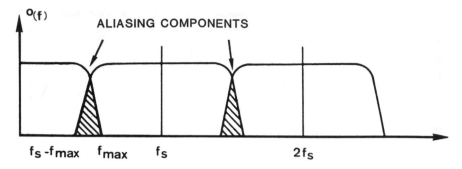

Figure 2.4 *If sampling frequency is too low, aliasing occurs*

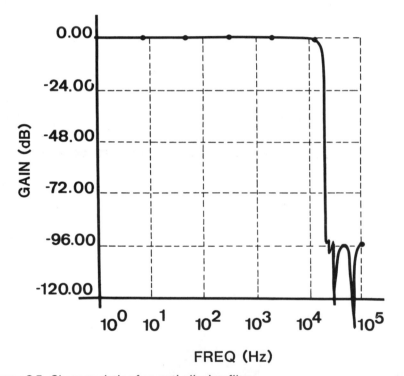

Figure 2.5 *Characteristic of an anti-aliasing filter*

Although sampling in Figure 2.3b seems much coarser than in Figure 2.3a, in both cases restitution of the original signal is perfectly possible.

Figure 2.2.3b also shows that f_s must be greater than $2f_{max}$ otherwise the original spectrum would overlap with the modulated part of the spectrum, and consequently be inseparable from it (Figure 2.4).

For example, a 20 kHz signal sampled at 35 kHz produces a 5 kHz difference frequency. This phenomenon is known as **aliasing**.

To avoid aliasing due to harmonics of the analog signal, a very sharp cut-off filter (known as an anti-aliasing filter) is used in the signal path to remove harmonics before sampling takes place. Characteristic of a typical anti-aliasing filter is shown in Figure 2.5.

Sampling Frequency

From this, it is easy to understand that the selection of the sampling frequency is very important. On one hand, selecting too high a sampling rate would increase the hardware costs dramatically. On the other hand, since ideal low-pass filters do not exist, a certain safety margin must be incorporated in order to avoid any frequency higher than $\frac{1}{2}f_s$ passing through the filter with insufficient attenuation.

EIAJ format

For an audio signal with a typical bandwidth of 20 Hz to 20 kHz, the lowest sampling frequency which corresponds to the Nyquist theorem is 40 kHz. At this frequency a very steep and, consequently, very expensive anti-aliasing filter, is required. Therefore a sampling frequency of approximately 44 kHz is typically used, allowing use of an economical 'anti-aliasing filter'; flat until 20 kHz but with sufficient attenuation (60 dB) at 22 kHz to make possible aliasing components inaudible.

Furthermore, because the first commercially available digital audio recorders stored the digital signal using a standard helical scan video recorder, there had to be a fixed relationship between sampling frequency (f_s) and horizontal video frequency (f_h), so these frequencies could be derived from the same master clock by frequency division.

For the NTSC 525-line television system, a sampling frequency of 44,055944...Hz was selected, whereas for the PAL 625-line system, a frequency of 44,100 Hz was chosen. The difference between these two frequencies is only 0.1%, which is negligible for normal use (the difference translates as a pitch difference at playback, and 0.1% is entirely imperceptible).

Compact disc sampling rate

For compact disc, the same sampling rate as used in the PCM-F1 format, i.e., 44.1 kHz, was commonly agreed upon by the establishers of the standard.

Video 8 PCM

The video 8 recording standard also has a provision for PCM recording. The PCM data is recorded in a time-compressed form, into a 30° tape section of

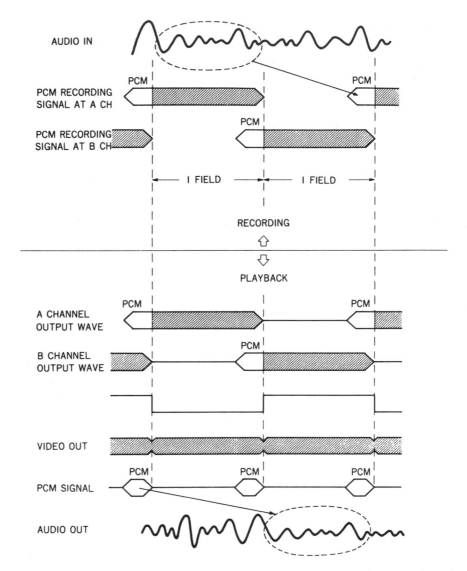

Figure 2.6 *Showing how PCM coded audio signals are recorded on a section of tape in a Video 8 track*

each video channel track (Figure 2.6). However, sampling frequency must be reduced to allow the data to fit.

Nevertheless, as the sampling frequency is exactly twice the video horizontal frequency (f_h), the audio frequency range is still more than 15 kHz, which is still acceptable for hi-fi recording purposes. Table 2.1 lists the horizontal frequencies for PAL and NTSC television standards, giving resultant sampling frequencies.

Also, since only 30° of the track are used for the audio recording, a multi-channel (6 channels) PCM recording is possible when the whole video recording area is used for PCM.

Table 2.1 *Horizontal and sampling frequencies of Video 8 recorders, on PAL and NTSC television standards*

	PAL	NTSC
f_h	15.625	15.734365
f_s	31.250	31.468530

Sampling rate for professional application

A sampling frequency of 32 kHz has been chosen by the EBU for PCM communications for broadcast and telephone, since an audio frequency range of up to 15 kHz is considered to be adequate for television and FM radio broadcast transmissions.

As we saw in Chapter 1, stationary-head digital audio recorders are most suited to giving a multi-track recording capability in the studio. One of the aspects of such studio recorders is that the tape speed must be adjustable, in order to allow easy synchronization between several machines and correct tuning. Considering a speed tuning range of, say, 10%, a sampling frequency of 44 kHz could decrease to less than 40 kHz, which is too low to comply with the Nyquist theorem. Therefore, such machines should use a higher sampling rate which at the lowest speed must still be above $2f_s$.

After a study of all aspects of this matter, 48 kHz has been selected as recommended sampling frequency for studio recorders. This frequency is compatible with television and motion-picture system frame frequencies (50 and 60 Hz) and has an integer relationship (3/2) with the 32 kHz sampling of the PCM network used by broadcast companies. The relationships between sampling frequency and frame frequencies (1/960, 1/800) enable the application of time coding, which is essential for editing and synchronization of the tape recorder. As there is no fixed relationship between 44.1 kHz (CD sampling frequency) and 48 kHz, Sony's studio recorders can use either frequency.

The reason why so much importance was attached to the integer relation of sampling frequencies is conversion: it is then possible to economically dub or convert the signals in the digital mode without any deterioration. Newly developed sampling rate converters, however, allow conversion also between non-related sampling rates, so that this issue has become less important than it used to be.

Sampling rates for R-DAT and S-DAT formats

In the DAT specification, multiple sampling rates are possible to enable different functions:

- 48 kHz, for highest-quality recording and playback
- 32 kHz, for high-quality recording with longer recording time or for recording on four tracks and for direct (digital) recording of digital broadcasts
- 44.1 kHz playback-only, for reproduction of commercial albums released on R-DAT or S-DAT cassettes.

Sample-Hold Circuits

In the practice of analog-to-digital conversion, the sampling operation is performed by **sample-hold circuits**, that store the sampled analog voltage for a time; during which the voltage can be converted by the A/D converter into a digital code. The principle of a sample-hold circuit is relatively simple and is shown in Figure 2.7.

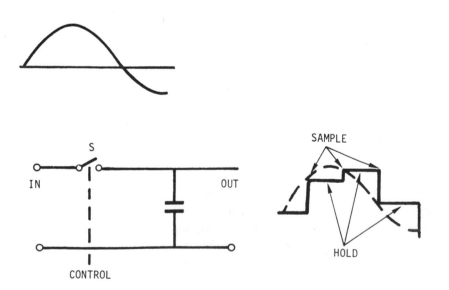

Figure 2.7 *Basic sample-hold circuit*

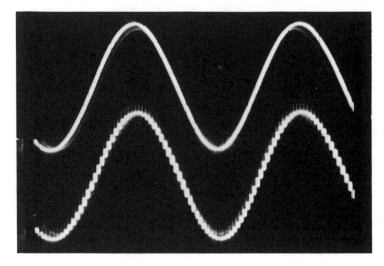

Figure 2.8 *Sine wave before (top) and after (bottom) sample-hold*

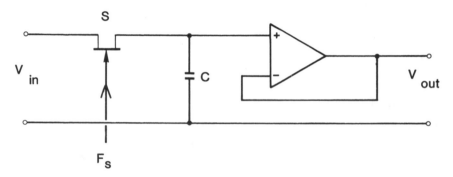

Figure 2.9 *FET input sample-hold circuit*

A basic sample-hold circuit is a 'voltage memory' device that stores a given voltage in a high-quality capacitor. To sample the input voltage, switch S closes momentarily: when S re-opens, capacitor C holds the voltage until S closes again to pass the next sample. Figure 2.8 is a photograph showing a sine wave at the input (top) and the output (bottom) of a sample-hold circuit.

Practical circuits have buffer amplifiers at input, in order not to load the source, and output, to be able to drive a load such as an A/D converter. The output buffer amplifier must have a very high input impedance, and very low bias current, so that the charge of the hold capacitor does not leak away. Also, the switch must be very fast and have low off-stage leakage. An actual sample hold circuit may use an analog (JFET-switch) and a high-quality capacitor, followed by a buffer amplifier (voltage-follower), as shown in Figure 2.9.

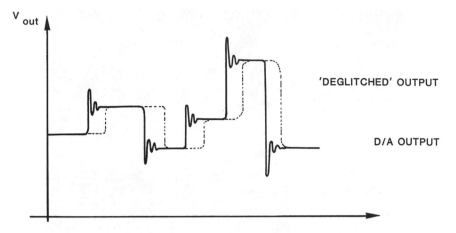

Figure 2.10 *The effect of using a sample-hold circuit as a deglitcher*

Sample-hold circuits are not only used in A/D conversion but also in D/A conversion, to remove transients (glitches) from the output of the D/A converter. In this case, a sample-hold circuit is often called a **deglitcher** (Figure 2.10).

Aperture Control

The output signal of a sampling process is in fact a pulse amplitude modulated (PAM) signal. It can be shown that, for sinusoidal input signals, the frequency characteristic of the sampled output is:

$$H(\omega_v) = \frac{t_o}{t_s} \cdot \frac{\sin \frac{t_o}{2}\omega_v}{\frac{t_o}{2}\omega_v}$$

In which ω_v = angular velocity of input signal $(= 2f_v)$
t_o = pulse width of the sampling pulse
t_s = sampling period.

At the output of a sample/hold circuit or a D/A converter, however, $t_o = t_s$. Consequently:

$$H(\omega_v)_{t_o = t_s} = \frac{\sin \frac{t_s}{2}\omega_v}{\frac{t_s}{2}\omega_v}$$

This means that at maximum admissible input frequency (which is half the sampling frequency), $\omega_v = \dfrac{\pi}{t_s}$, and consequently:

$$H\left(\frac{\pi}{t_s}\right)_{t_o\,=\,t_s} = \frac{\sin\dfrac{\pi}{2}}{\dfrac{\pi}{2}} \approx 0.64$$

This decreased frequency response can be corrected by an **aperature circuit,** which decreases t_o and restores a normal PAM signal (Figure 2.11).

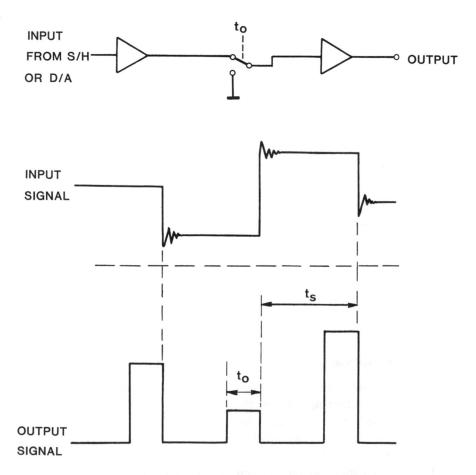

Figure 2.11 *Basic circuit and waveforms of aperture control circuit*

In most practical circuits $t_o = \dfrac{t_s}{4}$, which leads to:

$$H\left(\frac{\pi}{t_s}\right)_{ts} = \frac{t_s}{4} = \frac{\sin\dfrac{\pi}{8}}{\dfrac{\pi}{8}} \approx 0.97$$

This is an acceptable value: reducing t_o further would also reduce the average output voltage too much and thus worsen the signal-to-noise ratio.

Figure 2.12 shows the frequency response for some values of t_o.

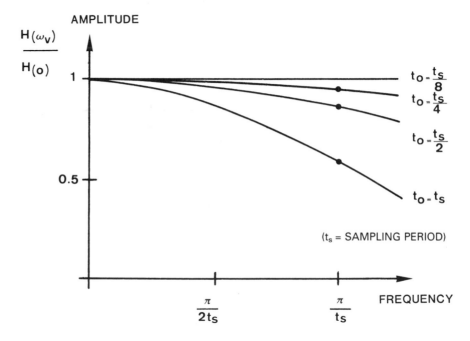

Figure 2.12 *Showing output characteristic of an aperture control circuit, as functions of aperture time and frequency response*

Characteristics and Terminology of Sample-Hold Circuits

In a sample-hold circuit, the **accuracy** of the output voltage depends on the quality of the buffer amplifiers, on the leakage current of the holding capacitor and of the sampling switch. Unavoidable leakage generally causes the output voltage to decrease slightly during the 'hold' period, in a process known as **droop**.

In fast applications, **acquisition time** and **settling time** are also important. Acquisition time is the time needed after the transition from hold to sample periods for the output voltage to match the input, within a certain error band. Settling time is the time needed after the transition from sample to hold periods to obtain a stable output voltage. Both times obviously define the maximum sampling rate of the unit.

Aperture time is the time interval between beginning and end of the transition from sample to hold periods; also terms like **aperture uncertainty** and **aperture jitter** are used to indicate variations in the aperture time and consequently variations of the sample instant itself.

3
Principles of Quantization

Even after sampling the signal is still in the analog domain: the amplitude of each sample can vary *infinitely* between analog voltage limits. The decisive step to the digital domain is now taken by **quantization** (see Figure 3.1), i.e.,

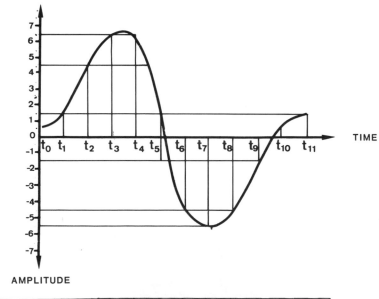

Sample	t_0	t_1	t_2	t_3	t_4	t_5	t_6	t_7	t_8	t_9	t_{10}	t_{11}
Value	1	2	5	7	6	0	5	6	5	2	1	1
4–bit code (2's complement)	0001	0010	0101	0111	0110	0000	1011	1010	1011	1110	0001	0001

Figure 3.1 *Principle of quantization*

replacing the *infinite* number of voltages by a *finite* number of corresponding values.

In a practical system the analog signal range is divided into a number of regions (in our example, 16), and the samples of the signal are assigned a certain value (say, -8 to $+7$) according to the region in which they fall. The values are denoted by digital (binary) numbers. In Figure 3.1, the 16 values are denoted by a 4-bit binary number, as $2^4 = 16$.

The example shows a **bipolar** system in which the input voltage can be either positive or negative (the normal case for audio). In this case, the coding system is often the **2's complement** code, in which positive numbers are indicated by the natural binary code while negative numbers are indicated by complementing the positive codes (i.e., changing the state of all bits) and adding one. In such a system, the most significant bit (MSB) is used as a **sign bit**, and is 'zero' for positive values but 'one' for negative values.

The regions into which the signal range is diverted are called **quantization intervals**, sometimes represented by the letter Q. A series of n bits represent-

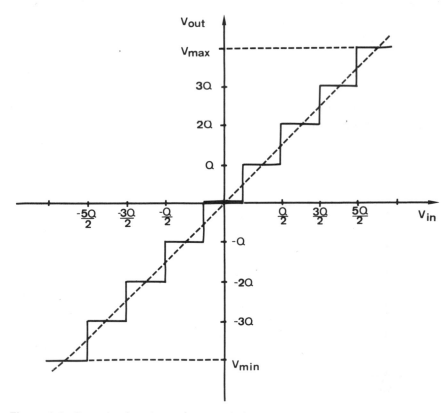

Figure 3.2 *Quantization stage characteristic*

ing the voltage corresponding to a quantization interval is called a **word**. In our simple example a word consists of four bits. Figure 3.2 shows a typical quantization stage characteristic.

In fact, quantization can be regarded as a mechanism in which some information is thrown away, keeping only as much as necessary to retain a required accuracy (or fidelity) in an application.

Quantization Error

By definition, because all voltages in a certain quantization interval are represented by the voltage at the *centre* of this interval, the process of quantization is a non-linear process and creates an error, called **quantization error** (or, sometimes, **round-off error**). The maximum quantization error is obviously equal to half the quantization interval Q, except in the case that the input voltage widely exceeds the maximum quantization levels ($+$ or $-V_{max}$), when the signal will be rounded to these values. Generally, however, such overflows or underflows are avoided by careful scaling of the input signal.

So, in the general case, we can say that:

$$-Q/2 < e_{(n)} < Q/2$$

where $e_{(n)}$ is the quantization error for a given sample n.

It can be shown that, with most types of input signals, the quantization errors for the several samples will be *randomly* distributed between these two limits, or in other words, its probability density function is flat (Figure 3.3).

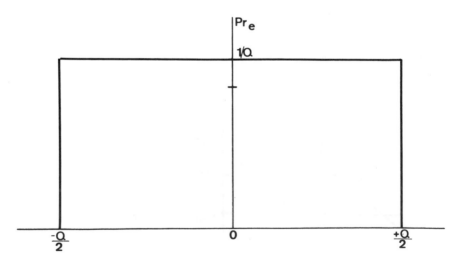

Figure 3.3 *Probability density function of quantization error*

There is a very good analogy between quantization error in digital systems and noise in analog systems: one can indeed consider the quantized signal as a perfect signal plus quantization error (just like an analog signal can be considered to be the sum of the signal without noise *plus* a noise signal) (see Figure 3.4). In this manner, the quantization error is often called **quantization noise**, and a 'signal-to-quantization noise' ratio can be calculated.

Calculation of Theoretical Signal-to-Noise Ratio

In an n-bit system, the number of quantization intervals N can be expressed as:

$$N = 2^n \tag{1}$$

If the maximum amplitude of the signal is V, the quantization interval Q can be expressed as:

$$Q = \frac{V}{N-1} \tag{2}$$

As the quantization noise is equally distributed within $+/- Q/2$, the quantization noise power N_a is:

$$N_a = \frac{2}{Q} \int_0^{Q/2} x^2 \, dx = \frac{2}{Q} \left(\frac{(Q/2)^3}{3} \right) = \frac{1}{12} Q^2 \tag{3}$$

If we consider a sinusoidal input signal with p-p amplitude V, the signal power is:

$$S = \frac{1}{2} \int_0^2 \left(\frac{2}{Q} \sin x \right)^2 dx = \frac{1}{8} V^2 \tag{4}$$

Consequently, the power ratio of signal-to-quantization noise is:

$$\frac{S}{N_a} = \frac{V^2/8}{Q^2/12} = \frac{V^2/8}{V^2/(N-1)^2 \cdot 12} \approx \frac{3}{2} N^2 \text{ (for } N \gg 1) \tag{5}$$

Or by substituting equation (1):

$$\frac{S}{N_a} = \frac{3}{2} \left(2^{2n} \right) = 3 \left(2^{2n-1} \right)$$

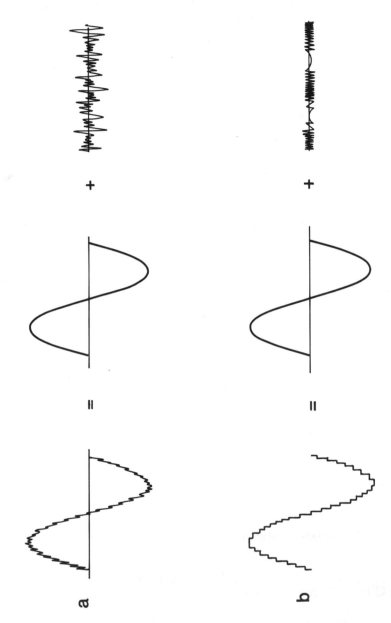

Figure 3.4 *Analogy between quantization error and noise*

Table 3.1 *Truth table for 16-bit 2's complement binary system*

```
0 1 1 1 1 1 1 1 1 1 1 1 1 1 1 1 = +32767
0 1 1 1 1 1 1 1 1 1 1 1 1 1 1 0 = +32766
0 1 1 1 1 1 1 1 1 1 1 1 1 1 0 1 = +32765
0 1 1 1 1 1 1 1 1 1 1 1 1 1 0 0 = +32764
                    :
                    :
                    :
0 0 0 0 0 0 0 0 0 0 0 0 0 0 1 1 = +3
0 0 0 0 0 0 0 0 0 0 0 0 0 0 1 0 = +2
0 0 0 0 0 0 0 0 0 0 0 0 0 0 0 1 = +1
0 0 0 0 0 0 0 0 0 0 0 0 0 0 0 0 =  0
1 1 1 1 1 1 1 1 1 1 1 1 1 1 1 1 = −1
1 1 1 1 1 1 1 1 1 1 1 1 1 1 1 0 = −2
1 1 1 1 1 1 1 1 1 1 1 1 1 1 0 1 = −3
1 1 1 1 1 1 1 1 1 1 1 1 1 1 0 0 = −4
                    :
                    :
                    :
1 0 0 0 0 0 0 0 0 0 0 0 0 0 1 1 = −32765
1 0 0 0 0 0 0 0 0 0 0 0 0 0 1 0 = −32766
1 0 0 0 0 0 0 0 0 0 0 0 0 0 0 1 = −32767
1 0 0 0 0 0 0 0 0 0 0 0 0 0 0 0 = −32768
```

Expressed in decibels, this gives:

$$\text{S/N (dB)} = 10 \log (\text{S/N}_a) = 10 \log 3(2^{2n-1})$$

Working this out gives:

$$\text{S/N (dB)} = 6.02 \times n + 1.76$$

A 16-bit system, therefore, gives a theoretical signal-to-noise ratio of 98 dB; a 14-bit system gives 86 dB.

In a 16-bit system, the digital signal can take 2^{16} (i.e., 65,535) different values, according to the truth table shown in Table 3.1.

Masking of Quantization Noise

Although, generally speaking, the quantization error is randomly distributed between + and −Q/2 (see Figure 3.1) and is consequently similar to analog white noise, there are some cases in which it may become much more noticeable than the theoretical signal-to-noise figures would indicate.

The reason is mainly that, under certain conditions, quantization can create harmonics in the audio passband which are not directly related to the input signal; and audibility of such distortion components is much higher than in the 'classical' analog cases of distortion. Such distortion is known as **granulation noise**, and, in bad cases, it may become audible as beat tones.

Auditory tests have shown that to make granulation noise just as perceptible as 'analog' white noise, the measured signal-to-noise ratio should be up to 10–12 dB higher. To reduce this audibility, there are two possibilities:

a) to increase the number of bits sufficiently (which is very expensive)
b) to 'mask' the digital noise by a small amount of analog white noise, known as **'dither noise'**.

Although such an addition of 'dither noise' actually worsens the overall signal-to-noise ratio by several dB, the highly audible granulation effect can be very effectively masked by it. The technique of adding 'dither' is well known in the digital signal processing field; particularly in video applications, where it is used to reduce the visibility of the noise in digitized video signals.

Conversion Codes

In principle, any digital coding system can be adopted to indicate the different analog levels in A/D or D/A conversion, as long as they are properly defined. Some, however, are better for certain applications than others. Two main groups exist: unipolar codes and bipolar codes. Bipolar codes give information on both the magnitude and the sign of the signal, which makes them preferable for audio applications.

Unipolar codes

Depending upon application, the following codes are popular:

Natural binary code
The MSB has a weight of 0.5 (i.e., 2^{-1}), the second bit has a weight of 0.25 (2^{-2}), and so on, until the least significant bit (LSB) which has a weight of 2^{-n}. Consequently, the maximum value that can be expressed (when all bits are one) is $1 - 2^{-n}$, i.e., full-scale minus one LSB.

BCD code
The well-known 4-bit code in which the maximum value is 1001 (decimal 9), after which the code is reset to 0000. A number of such 4-bit codes is combined in case we want, for instance, a direct read-out on a numeric scale such as in digital voltmeters. Because of this maximum of ten levels, this code is not used for audio purposes.

Gray code

Used when the advantage of changing only one bit per transition is important, for instance in position encoders where inaccuracies might otherwise give completely erroneous codes. It is easily convertible to binary. Not used for audio.

Bipolar codes

These codes are similar to the unipolar natural binary code, but one additional bit, the sign bit, is added. Structure of the most popular codes is compared in Table 3.2.

Sign magnitude

The magnitude of the voltage is expressed by its normal (unipolar) binary code, and a sign bit is simply added to express polarity.

An advantage is that the transition around zero is simple; however, it is more difficult to process and there are two codes for zero.

Offset binary

This is a natural binary code, but with zero at minus full scale; this makes it relatively easy to implement and to process.

Two's complement

Very similar to offset binary, but with the sign bit inverted. Arithmetically, a two's complement code word is formed by complementing the positive value and adding 1 LSB. For example:

$+2 = 0010$
$-2 = 1101 + 1 = 1110$

It is a very easy code to process, for instance, positive and negative numbers added together always give zero (disregarding the extra carry). For example:

```
  0010
+1110
 ─────
  0000
```

It is the code almost universally used for digital audio; there is however (as with offset binary), a rather big transition at zero – all bits change from 1 to 0.

One's complement

Here negative values are full complements of positive values. This code is not commonly used.

Table 3.2 *Common bipolar codes*

Decimal fraction Number	Positive reference	Negative reference	Sign + magnitude	Two's complement	Offset binary	One's complement
+7	$+\frac{7}{8}$	$-\frac{7}{8}$	0 1 1 1	0 1 1 1	1 1 1 1	0 1 1 1
+6	$+\frac{6}{8}$	$-\frac{6}{8}$	0 1 1 0	0 1 1 0	1 1 1 0	0 1 1 0
+5	$+\frac{5}{8}$	$-\frac{5}{8}$	0 1 0 1	0 1 0 1	1 1 0 1	0 1 0 1
+4	$+\frac{4}{8}$	$-\frac{4}{8}$	0 1 0 0	0 1 0 0	1 1 0 0	0 1 0 0
+3	$+\frac{3}{8}$	$-\frac{3}{8}$	0 0 1 1	0 0 1 1	1 0 1 1	0 0 1 1
+2	$+\frac{2}{8}$	$-\frac{2}{8}$	0 0 1 0	0 0 1 0	1 0 1 0	0 0 1 0
+1	$+\frac{1}{8}$	$-\frac{1}{8}$	0 0 0 1	0 0 0 1	1 0 0 1	0 0 0 1
0	0+	0−	0 0 0 0	0 0 0 0	1 0 0 0	0 0 0 0
0	0−	0+	1 0 0 0	(0 0 0 0)	(1 0 0 0)	1 1 1 1

1110	0111	1111	1001	$+\frac{1}{8}$	$-\frac{1}{8}$	-1
1101	0110	1110	1010	$+\frac{2}{8}$	$-\frac{2}{8}$	-2
1100	0101	1101	1011	$+\frac{3}{8}$	$-\frac{3}{8}$	-3
1011	0100	1100	1100	$+\frac{4}{8}$	$-\frac{4}{8}$	-4
1010	0011	1011	1101	$+\frac{5}{8}$	$-\frac{5}{8}$	-5
1001	0010	1010	1110	$+\frac{6}{8}$	$-\frac{6}{8}$	-6
1000	0001	1001	1111	$+\frac{7}{8}$	$-\frac{7}{8}$	-7
	(0000)	(1000)		$+\frac{8}{8}$	$-\frac{8}{8}$	-8

4
Overview of A/D Conversion Systems

Linear (or Uniform) Quantization

In all the examples seen so far, the quantization intervals Q were identical. Such quantization systems are commonly termed **linear** or **uniform**. Regarding simplicity and quality, linear systems are certainly best. However, linear systems are rather costly in terms of required bandwidth and conversion accuracy. Indeed, a 16-bit audio channel with a sampling frequency of 44.056 kHz gives a bit stream of at least $16 \times 44.056 = 705 \times 10^3$ bit s^{-1}, which requires a bandwidth of 350 kHz – 17.5 times the bandwidth of the original signal. In practice, a wider bandwidth than this is required because more bits are needed for synchronization, error correction and other purposes.

Since the beginning of PCM telephony, ways to reduce the bandwidths that digitized audio signals require have been developed. Most of these techniques can also be used for digital audio.

Companding Systems

If, in a quantizer, the quantization intervals Q are not identical, we talk about **non-linear quantization**. It is, for instance, perfectly possible to change the quantization intervals according to the level of the input signal. In general, in such systems, small-level signals will be quantized with more closely spaced intervals, while larger signals can be quantized with bigger quantization intervals. This is possible because the larger signals more or less mask the unavoidably higher noise levels of the coarser quantization.

Such a non-linear quantization system can be thought to consist of a linear system, to which a compander has been added. In such a system, the input signal is first compressed, following some non-linear law F(x), then linearly

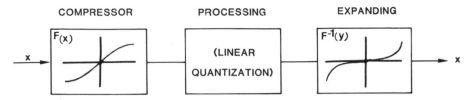

Figure 4.1 *Non-linear quantization*

quantized, processed and then after reconversion, expanded by the reverse non-linearity $F^{-1}(y)$ (see Figure 4.1). The overall effect is analogous to companders used in the analog field (e.g., Dolby, dBx and others).

The non-linear laws which compressors follow can be shown in graphical forms as curves. One compressor curve used extensively in digital telephony in North America for the digitization of speech, is the 'μ-law' curve. This curve is characterized by the formula:

$$F(x) = V \frac{V_{\log}(1 + \mu x/V)}{\log(1 + \mu)}$$

Curves for this equation are shown in Figure 4.2 for several values of μ. In Europe, the 'A law' curve is more generally used (Figure 4.3).

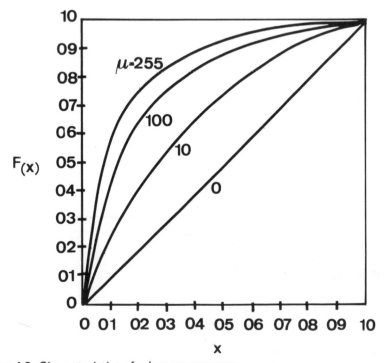

Figure 4.2 *Characteristics of μ-law compressors*

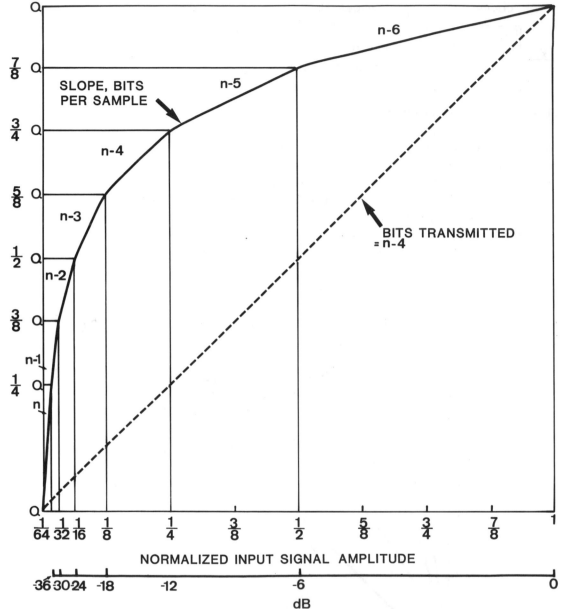

Figure 4.3 *A-law characteristic curve*

The (dual) formula for the 'A law' is:

$$F(x) = Ax/1 + \log A \text{ for } 0 < x < V/a$$

$$F(x) = V + V \log (Ax/V)/1 + \log A \text{ for } V/a < x < V$$

In practice, it is important that the non-linearities at the input and the output of any audio system are very closely matched. This is difficult to achieve with analog techniques so compressors are usually built in to the conversion process.

The big advantage of these companded systems is that signal-to-noise ratio becomes less dependent on the level of the input signal; the disadvantage, however, is that the noise level follows the level of the signal, which may lead to audible noise modulation.

Floating-Point Conversion

A special case of non-linear quantization, used in professional audio systems, is the 'floating-point converter' (Figure 4.4).

Sampled signal is sent through several selectable paths, each with a different gain; depending on the input level of the signal. Path, and hence gain, is selected by a logic monitor circuit in order to make maximum use of the linear A/D converter without overloading it. The output from the A/D converter, called 'mantissa' in an analogy with logarithmic annotation, is meaningless without a way to indicate the gain that was originally selected. This information is provided by a logic output from the monitor circuit, called 'exponent'. Exponent and mantissa, taken together, give an unambiguous digital word that can be reconverted to the original signal by selecting the corresponding (inverse) gains in the decoding stage. In this way two bits of exponent can indicate four different gains. If we select these gains as 0, 6, 12 and 18 dB for instance, the two additional bits provide an increase of 18 dB over the dynamic range of the basic system.

Because the signal level determines basic system gain, noise modulation is unavoidable. This may become audible, for instance, with a high-level,

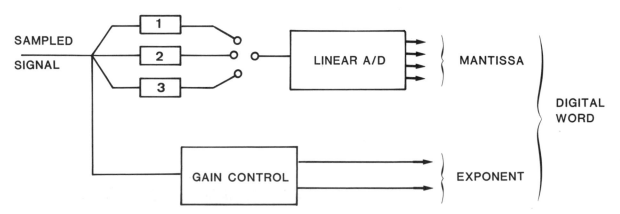

Figure 4.4 *Floating-point converter principle*

low-frequency, signal: in this case, noise modulation will not be masked by the signal.

Due to effects of noise modulation, a distinction must be made between the dynamic range and the signal-to-noise ratio. The dynamic range can be defined as:

$$\frac{\text{maximum signal level (RMS)}}{\text{RMS level of quantization noise } without \text{ signal}}$$

whereas the signal-to-noise ratio is:

$$\frac{\text{signal level (RMS)}}{\text{RMS level of quantization noise } with \text{ signal}}$$

A curve for the signal-to-noise ratio of a typical floating-point converter with a 10-bit mantissa, a 3-bit exponent and 6 dB gain steps is shown in Figure 4.5. Although, theoretically, this system provides the same dynamic range as a 17-bit linear system (i.e., over 100 dB), the signal-to-noise ratio is unacceptable for high-quality purposes.

In spite of this, high-quality floating-point converters having, say, a 13-bit mantissa and 3-bit exponent are still considered for digital audio purposes, as they are considerably cheaper than linear systems.

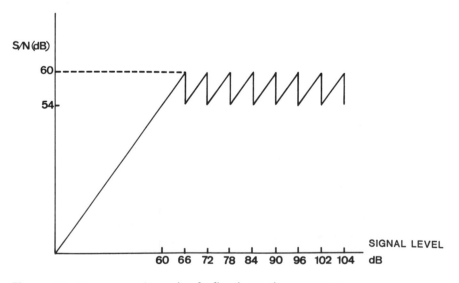

Figure 4.5 *Signal-to-noise ratio of a floating-point converter*

Block Floating-Point Conversion

When a low bandwidth is of utmost importance, **block conversion** can be used. This technique is also known as **near-instantaneous companding** (in contrast to basic floating point or other companding systems). The term 'near-instantaneous' is used to describe the fact that not every sample is scaled by an exponent, but a number of successive samples (usually 32). Each block of samples is then followed by a scale factor word, so that, at the receiving end, each block can be correctly scaled up again (Figure 4.6).

This system is rather expensive as far as hardware is concerned, but permits significant reductions in bit rates. Consequently, a typical application is digital transmission of audio signals in radio networks.

Subjective listening tests have shown that an original 14-bit system compressed to 10 bits is almost indistinguishable from a 13-bit linear system, although the signal-to-noise ratio limitations of a floating-point converter remain valid.

An example of such a system is the BBC's NICAM-3 (near-instantaneous companding audio multiplex) which permits transmission of six audio channels over one (standard) telephony 2048 kbits/s circuit.

Differential PCM and Delta Modulation

Instead of transmitting the exact binary value of each sample, it is possible to transmit only the difference between the current sample and the previous one. As this difference is generally small, a smaller number of bits can be used with no apparent degradation in performance. Operation is fairly straightforward: one sample is stored for the complete sample period, then added to

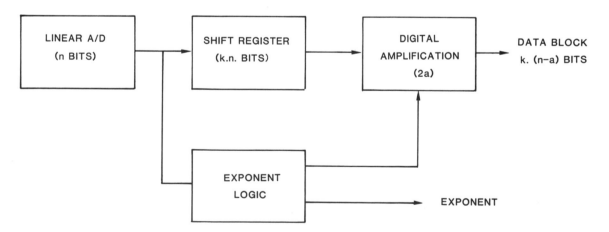

Figure 4.6 *A block floating-point converter*

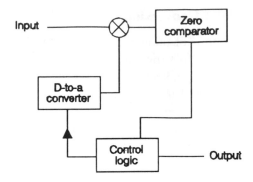

Figure 4.7 *1-bit A/D converter*

the received difference signal to obtain the next sample. This sample is then stored until the next received difference signal.

Differential PCM, in fact, is a special type of **predictive encoding**. In such encoding schemes, a prediction is generated for the current sample, based upon past data; the correcting signal is simply the difference between the prediction and the actual signal.

As sampling rate increases, the differences between previous and present samples become smaller, so that, in the extreme for very high sampling rates, only 1 bit is needed for the error signal to indicate the sign of the error; in this case we talk about **delta modulation**.

Figure 4.7 shows a basic single-bit A/D converter. The input signal is compared with the output of a 1-bit D/A converter, the resulting voltage is then compared with a reference and the output used to increment or decrement the DAC value. For any input signal the system needs to perform a certain number of iterations to obtain the required resolution. Each iteration results in a high or low signal at the output of the A/D converter. Looking at the output we see a pulse train whose mean value equals the level of the input signal. The analog input has been converted to a binary bit stream. The typical sampling frequency in 1-bit converters is several MHz.

Because the serial bit stream is of little practical use, it is mostly converted to a multibit format (e.g. 16 bit) with a much lower sampling rate. This is done in a digital filter, a so called decimation filter which includes noise shaping (Figure 4.8).

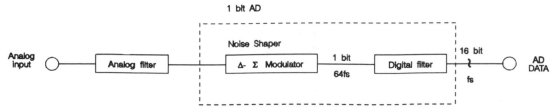

Figure 4.8 *Block diagram of 1-bit A/D converter*

In a further step, the transmitted data can be used to indicate not only the sign of the error, but also the step size. For example, a continuous series of ones means that the signal is quickly increasing, so the step size can be increased; if ones and zeros are alternating, step size can be reduced. Such strategies are called **adaptive differential PCM** (the quantization interval is changed) or **adaptive delta modulation** (the step size is changed). Although these techniques have some interesting theoretical and practical properties, it is presently difficult to use them for high-quality applications.

Super Bit Mapping (SBM)

SBM is an enhanced A/D conversion technique used for disc mastering and also on some recent DAT recorders (as from 1994). When the CD technology started, the amount of 16 bits per sample was specified. At that moment 16 bits seemed ambitious, as the available technology only allowed about 14–15 bits.

Now, however, it is no problem to convert substantially more than 16 bits per sample, and A/D conversion in professional sound studios is performed at 18 or 20 bits, if not higher. Of course, if a conversion is made at 18 or 20 bits and the disc specification allows 'only' 16 bits, there is a need for reduction. SBM uses the extra bits to increase the accuracy of the 16 bit signal.

A very important point to note is that the use of such techniques poses no compatibility problems at all. A disc to which the SBM principle has been applied can be played back without any problem in any CD player. If a digital recording was made using a 20 bit A/D converter, the last 4 bits (bits 16–19) cannot be implemented in the 16 bit data stream on the disc, but these bits can be used to increase the accuracy of the least significant bits of the 16 bit samples and thereby maintain compatibility, decrease the noise level and increase the sound quality. Figure 4.9 illustrates the SBM effect.

SBM operation is highly complex and involves higher order mathematics which is beyond the scope of this publication. It should always be remembered, however, that these calculations are applied to digital audio data not to analog signals. Figure 4.10 presents an SBM block diagram.

The input is audio data (more than 16 bits per sample), each channel is calculated separately, but it is obvious that in order to maintain sound

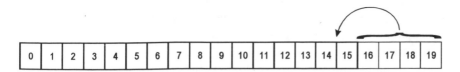

Figure 4.9 *SBM effect*

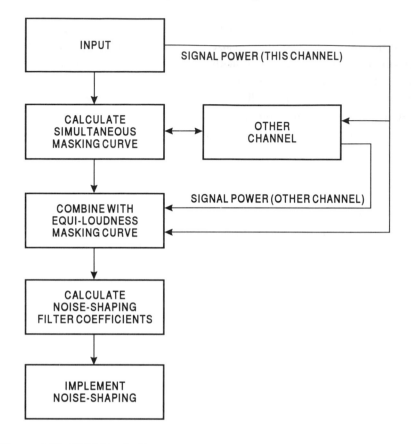

Figure 4.10 *SBM block diagram*

integrity, the calculations performed on one channel must have at some points input from the other channel.

Audio data will be calculated by blocks of 512 samples (512 samples at a sampling rate of 44.1 kHz equals a sound block of 11.6 ms).

Any DC component will be subtracted from the input.

The next block is the simultaneous masking curve calculation. The input blocks (512 samples left and right channel) are processed by Fast Fourier Transforms in order to analyse the input audio. This calculation determines the signal power in each frequency band, the total audio frequency spectrum being split into a number of well-defined sub-spectra. Note that this is a known psycho-acoustic fact: the human ear works according to similar principles: analysing signal power at predetermined subdivisions of the total audio spectrum.

At this point, and also in the next block, the equi-loudness masking block, the SBM system determines whether one part of the incoming signal can mask another part and where and at what level quantization noise is found.

The knowledge gained in previous blocks will be used to calculate filter coefficients which will be used to perform noise shaping. Noise shaping (see also Fig. 5.11) is a technology able to eliminate quantization noise contents from the audible spectrum.

In the next stage noise shaping is performed based upon all the previously performed calculations. Figure 4.11 illustrates the SBM noise shaping.

In comparison with noise shaping filters used at the D/A side of CD players it should be noted that in this case the filter coefficients are adapted continuously, so the operation is far more complex and needs a very high calculation speed.

The result of SBM is a 16 bit signal with a lower quantization noise, a higher linearity and therefore a better sound definition.

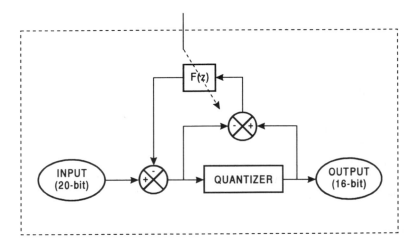

Figure 4.11 *SBM noise shaping*

5
Operation of A/D – D/A Converters

Some of the most important components in digital audio systems are the converters. Previous chapters have shown the need for high resolution to obtain a satisfactory signal-to-noise ratio.

In video applications, an 8-bit conversion is more than sufficient. A 14-bit conversion (or an equivalent) seems a minimum for good audio performance, and, for professional use, 16-bit conversion is required to leave a margin for further processing (e.g., filtering, mixing).

In the PCM-F1 and compact disc system, 16-bit converters are used, while the PCM video 8 system uses a 10-bit converter.

A/D Converters

Fundamentally, A/D converters operate in one of two general ways. They either: convert the analog input signal to a frequency or a set of pulses whose time is measured to provide a representative digital output, or: compare the input signal with a variable reference, using an internal D/A converter to obtain the digital output.

Basic types of A/D converters

Voltage-to-frequency, ramp, and integrating-ramp methods are the three leading conversion processes that use the time-measurement method. Successive approximation and parallel/modified parallel circuits rely on comparison methods.

Dual-slope integrating A/D converters

The dual-slope integrating A/D converter contains an integrator, some control logic, a clock, a comparator, and an output counter, as shown in

Figure 5.1. A graph of integrator output voltage against time is shown in Figure 5.2.

The input analog signal is initially switched to the integrator, and the output of the integrator ramps up for a time t_1. The slope of the ramp, and hence the integrator output voltage at the end of this time, depends on the amplitude of the analog input signal and the time constant τ of the integrator:

$$\frac{V_{IN}}{\tau}$$

So the integrator output voltage V_0 at the end of time t_1 is:

$$\frac{V_{IN}}{\tau} t_1$$

The reference signal is then switched to the integrator input, and the integrator output voltage ramps down until it returns to the starting voltage. The slope of the ramp during time t_2 similarly depends on the integrator time

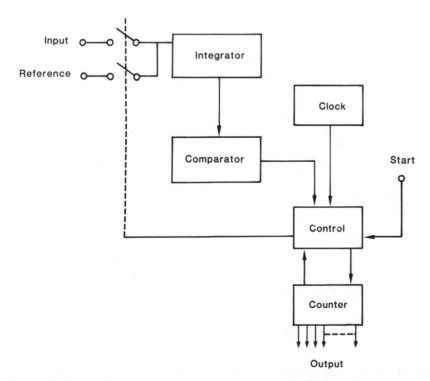

Figure 5.1 *Block diagram of a dual-slope integrating A/D converter*

constant and the integrator input voltage, this time the reference signal amplitude:

$$\frac{V_{REF}}{\tau}$$

So the integrator output voltage at the beginning time t_2 is:

$$\frac{V_{REF}}{\tau}t_2$$

But as these voltages are the same:

$$\frac{V_{IN}}{\tau}t_1 = \frac{V_{REF}}{\tau}t_2$$

Therefore:

$$\frac{V_{IN}}{V_{REF}} = \frac{t_2}{t_1}$$

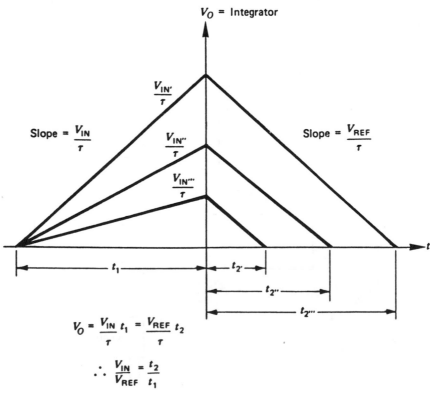

Figure 5.2 *Showing integrator output voltage as a function of time, during conversions*

which shows that time t_2 is totally dependent on the input signal amplitude, and independent of integrator time constant. By counting clock pulses during time t_2 a digital measure of the analog input signal's amplitude is made.

Average conversion time, i.e., time the converter takes to perform the conversion of an applied input signal, is two clock periods times the number of quantization levels. Thus, for a 12-bit converter with a 1 MHz clock, the average conversion time is: $2 \times 1\,\mu s \times 4096$ or 8.192 ms. The precise conversion time, however, depends on the applied input signal amplitude.

Due to this long conversion time integrating converters are not useful for digitizing high-speed, rapidly varying signals, although they are useful to 14-bit accuracy, offering high noise rejection and excellent stability with both time and temperature. They can be modified to increase conversion speeds and are used mostly in 8- to 12-bit converters for digital voltmeters (DVMs), digital panel meters (DPMs) and digital multimeters (DMMs). However, basic dual-slope integrating A/D converters are too slow for general computer applications.

Successive-approximation A/D converters

The main reasons that the successive-approximation technique is used almost universally in A/D conversion systems are: the reliability of the conversion technique, simplicity, and inherent high-speed data conversion. Conversion time is equal to the clock period times the number of bits being converted. Thus, for a 1 MHz clock, a 12-bit converter would take $12\,\mu s$ to convert an applied analog signal.

A successive approximation converter consists of a comparator, a register, control logic and a D/A converter. Output of the D/A converter is compared with the input analog voltage (Figure 5.3). Each bit line in the D/A converter corresponds to a bit position in the register. Initially, the converter is clear.

When an input signal is applied the control logic instructs the register to change its MSB to 1. This is changed by the D/A converter to an analog voltage equivalent to one-half the converter's full-scale range. If the input voltage is greater than this, the next most significant bit of the register becomes 1. If, however, the input is less, the next most significant bit remains 0. Then the circuit 'tries' the following bits through to the LSB, at which stage the conversion is complete. Thus the number of approximations occurring in any conversion equals the number of bits in the digital output. Figure 5.4 shows operation of the successive approximation A/D converter graphically.

Main advantage of the successive-approximation converter is speed and this is limited by the settling time of the DAC. Accuracy is limited by the accuracy of the DAC, and a high susceptibility to noise is its major drawback. As only one comparator is used and ancillary hardware is limited to logic, register and D/A converter, the successive-approximation technique provides an inexpensive A/D converter.

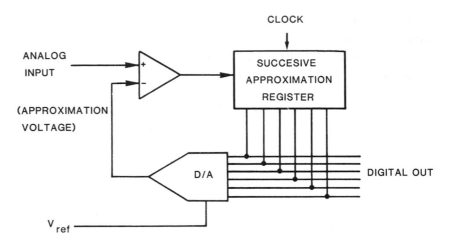

Figure 5.3 *Successive approximation A/D converter*

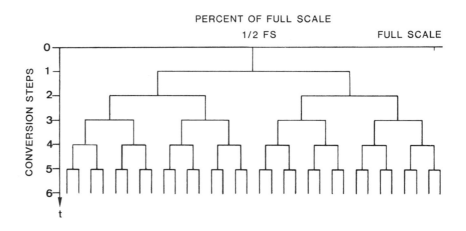

Figure 5.4 *Illustration of successive approximation conversion. The digitally generated voltage gets closer to the analog input voltage in a series of approximations; each approximation is half the preceding one*

Other types of A/D converters

Voltage-to-frequency converters

Figure 5.5 shows a typical voltage-to-frequency converter. Here, the input analog signal is integrated and fed to a comparator. When the comparator changes its state, the integrator is reset and the process repeats itself. The counter counts the number of integration cycles for a given time to provide a digital output.

The principal advantage of this type of conversion is its excellent noise

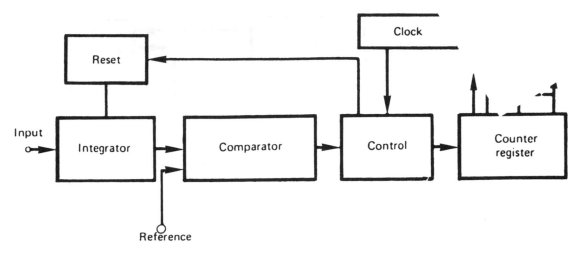

Figure 5.5 *Voltage-to-frequency A/D converter*

rejection due to the fact that the digital output represents the average value of the input signal. Voltage-to-frequency conversion, however, is too slow for use in data-acquisition system applications because it operates bit-serially (with a maximum of approximately 1000 conversions/s). Its applications are mostly in digital voltmeters (DVMs) using converters with resolutions of 10 bits or less.

Ramp converters
Ramp conversion works by continuously comparing a linear reference ramp signal with the input signal using a comparator (Figure 5.6). The comparator initiates a counter when changing state and the counter counts clock pulses during the time the comparator is logically HIGH; the count is therefore proportional to the magnitude of the input signal. The counter output is the digital representation of the analog input.

This method is slightly faster than the previous one, but it requires a highly linear ramp source in order to be effective. It does offer good 8- to 12-bit differential linearity for applications requiring high accuracy.

Parallel A/D converters
Parallel-series and straight parallel converters are used primarily where extremely high speed is required, taking advantage of the fact that the propagation time through a chain of amplifiers is equal to the square root of the number of stages times the individual setting time, as opposed to adding up the times of each stage. By adding a comparator for every binary-weighted network, as shown in Figure 5.7, it is possible to take advantage of this higher

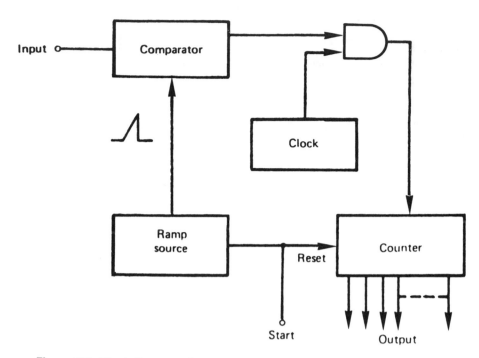

Figure 5.6 *Block diagram of a ramp converter*

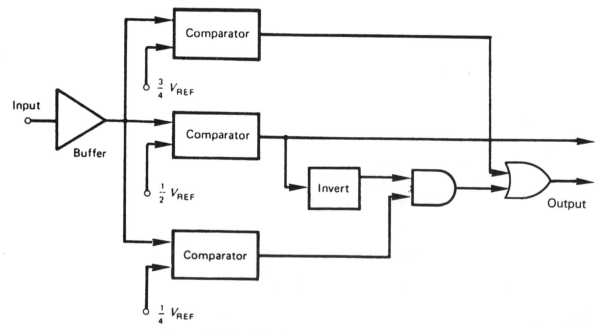

Figure 5.7 *Parallel A/D converter*

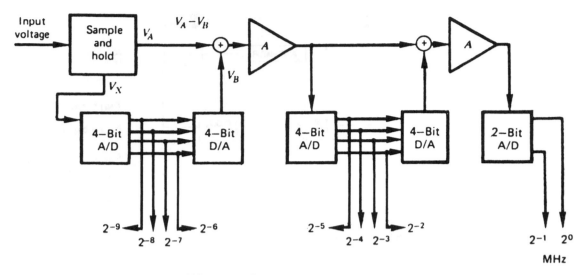

Figure 5.8 *Sequential parallel A/D conversion*

speed. Parallel A/D converters are often called **flash converters** because of their high operating speeds.

The parallel A/D converter of Figure 5.7 uses one comparator for each input quantization level (i.e., a 6-bit converter would have 6 comparators). Conversion is straightforward; all that is required besides the comparators is logic for decoding the comparator outputs.

Because only comparators and logic gates stand between the analog inputs and digital outputs, extremely high speeds of up to 50,000,000 samplings/s can be obtained at low resolutions of 6 bits or less. The fact that the number of comparator and logic elements increases with resolution obviously makes this converter increasingly impractical for resolutions greater than 6 bits.

Modified parallel designs can provide a good tradeoff between hardware complexity and the resolution/speed combination at a slight addition in hardware and a sacrifice in speed. They can provide up to 100,000 conversions/s for up to 14-bit resolutions. Sequential conversion (Figure 5.8), for example, is often used for such applications. However, because of the increase in the number of comparators and the need to use an amplifier for every weighting network, cost is considerably more than that of a successive approximation.

The first 4-bit converter in the circuit in Figure 5.8 provides the 4 most significant bits in parallel. These outputs are converted back to an analog voltage which is subtracted from the input. The difference is applied to the next converter and the process is continued until the required 10 bits are

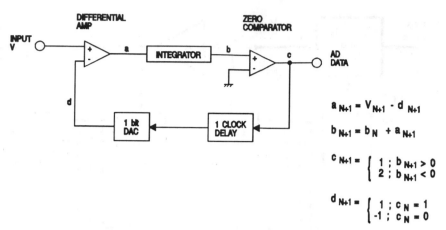

Figure 5.9 *First-order Delta-Sigma modulator*

obtained. This approach gives a reasonable tradeoff among speed, cost, and accuracy.

Delta-sigma modulator

A delta-sigma modulator is the key device in a 1 bit A/D converter. Figure 5.9 shows a first-order delta-sigma modulator. Operation is performed at each clock cycle, which corresponds to the oversampling frequency. At the beginning of each clock cycle, the differential amplifier outputs the difference between the input voltage V and the output voltage of the single-bit D/A converter. The integrator adds the voltage a to its own output from the preceding clock cycle. This voltage b is provided to the zero comparator. The output of the comparator will be logically HIGH or LOW, depending on voltage b being higher or lower than 0 V. The output then becomes a piece of single-bit A/D data, which is also used to determine the output of the 1 bit DAC for the next clock cycle. The 1 bit DAC outputs a positive full-scale voltage if its input is HIGH and a negative full-scale voltage if its input is LOW. Table 5.1 shows an example of actual operation in which the input is 0.6 V, with the full-scale voltage being ±1 V and all initial values 0.

The 1 bit A/D converter outputs only HIGH or LOW, which has no meaning in itself, this only becomes meaningful when a string of 1 bit data is averaged.

Because of the high sampling frequency (64 times oversampling) a very gentle low pass filter can be used, resulting in low phase distortion. Compared to successive approximation A/D converters single bit A/D converters provide better performance while circuit complexity and cost remain equal.

Table 5.1 *Operation example of Δ–Σ modulator*

Clock	d	a	b	c
0	0	0.6	0.6	1
1	1	−0.4	0.2	1
2	1	−0.4	−0.2	0
3	−1	1.6	1.4	1
4	1	−0.4	1.0	1
5	1	−0.4	0.6	1
6	1	−0.4	0.2	1
7	1	−0.4	−0.2	0
8	−1	1.6	1.4	1
9	1	−0.4	1.0	1
10	1	−0.4	1.0	1
11	1	−0.4	0.2	1
12	1	−0.4	−0.2	0
13	−1	1.6	1.4	1
14	1	−0.4	1.0	1
15	1	−0.4	0.6	1

0.6 V fixed input, ±1 V full-scale voltage

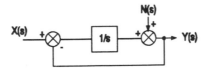

Figure 5.10 *Δ–Σ modulator or noise shaper*

Noise shaping
A delta-sigma modulator is sometimes also called a noise shaper because it passes signals and noise according to different transfer functions (Figure 5.10). The signal transfer function for the modulator simplifies to:

$$\frac{Y(s)}{X(s)} = \frac{1}{s+1}$$

This is the s-domain representation of a first-order low pass filter. Deriving the noise transfer function for the same modulator produces:

$$\frac{Y(s)}{N(s)} = \frac{s}{s+1}$$

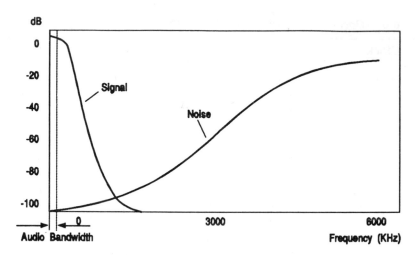

Figure 5.11 *Transfer functions of noise shaper*

This is the s-domain representation of a simple high-pass filter. Plotting the transfer functions gives the result shown in Figure 5.11. The signal is attenuated at higher frequencies, while the noise is shaped so that very little of its content is in the low frequency region. By using higher order delta-sigma modulators the in-band noise can even be reduced further, however out-of-band noise will increase. In practice a third or fourth order delta-sigma modulator is used to avoid stability problems while still using most of the noise shaping capabilities.

High Density Linear A/D Converter

The A/D converter currently used in Sony R-DAT recorders and some MiniDisc recorders is the high density linear converter. This converter uses two 4th order delta-sigma modulators for simultaneous sampling of two audio channels. Its output is a serial data signal with 16-bit resolution coded in 2's complement at a sampling frequency of 32 kHz, 44.1 kHz or 48 kHz. The delta-sigma modulators operate at 64 times oversampling, for a system sampling frequency of 48 kHz, the analog audio signals are actually sampled at:

$$f_s = 48\,\text{kHz} \times 64 = 3.072\,\text{MHz}$$

This high sampling frequency eliminates the need for a sample-hold circuit and a complex analog low-pass filter. A first order RC network can be used as anti-aliasing filter because the audio signals do not normally contain frequencies above $1/2\ f_s$ (approx. 1.5 MHz). As a consequence, circuit complexity is greatly reduced.

The 1-bit stream, output from the delta-sigma modulators, is hardly usable for further application. A built-in digital filter recalculates the single bit input data at $64\ f_s$ to 16-bit words at f_s. The so-called requantization is performed by the decimation filters.

Figure 5.12 shows the block diagram of the high density linear converter. This converter is clearly separated into an analog and a digital section. The analog part contains the voltage reference source and the delta-sigma modulators while the digital part includes the digital filters, a controller and the output interface. Each section has its own power supply to prevent digital noise from entering the analog signals.

During initialization the converter performs an automatic calibration to compensate for possible offset errors in the converter itself. The analog inputs are grounded while the resulting output is measured and its value stored in SRAM as an offset. After calibration, the actual data are corrected by this offset value before being output.

With a harmonic distortion (THD) of less than 0.002% and a signal-to-noise-ratio of more than 94 dB (EAIJ), the high density linear converter is ideally suited for high quality digital audio signal processing. This type of

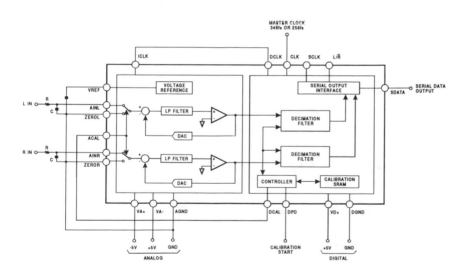

Figure 5.12 *High Density Linear Converter as used by Sony in digital audio equipment*

converter is used in all Sony's R-DAT recorders such as DTC-77ES and DTC-59ES, as well as in the MDS-101 MiniDisc recorder.

D/A Conversion in Digital Audio Equipment

Weighted current D/A converter

The most common D/A converter in digital audio is the weighted current type. It consists of a series of electronic switches, each of which is connected to a current source. In an n-bit D/A converter there are n weighted current sources with a current value of $2^{(n-1)}$ times the original current value l. The current sources corresponding to the weight of the bit in the digital data signal are added to obtain a current that represents the value of the digital data. In fact, the digital data input directly controls the electronic switches that turn the current sources on or off. A current-to-voltage amplifier converts the obtained current into a voltage before being output. Figure 5.13 shows a typical block diagram of the weighted current type or current summing type D/A converter.

Although this is basically a very simple converter, it has some serious drawbacks. The constant current sources must be very accurately matched to prevent non-linear distortion. Suppose that the current l equals 0.1 mA, the constant current source for the MSB should then deliver $2^{(n-1)} \times l = 3276.8$ mA or more than 3A! Moreover, to maintain 16-bit resolution the accuracy of the MSB constant current source must be better than the current delivered by the LSB constant current source.

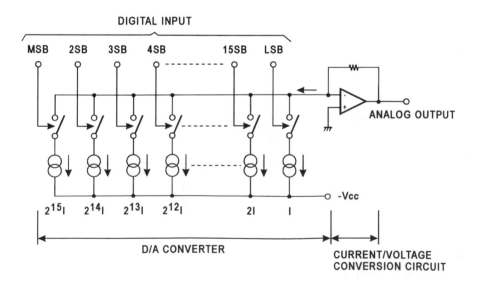

Figure 5.13 *Current summing type D/A converter*

Another type of D/A converter, more or less based on the same principles is the ladder network type D/A converter as shown in Fig. 5.14. The so-called 'ladder network' is composed of resistors with values R and 2R to form a voltage divider. The electronic switches, controlled by the digital input data, change the output voltage of the voltage divider. The output voltage therefore represents the digital input signal. Accuracy is also a problem because all the resistors must be perfectly matched to ensure linearity. Temperature variations and aging inevitably have a bad influence on the D/A converter linearity. The ON-resistance of the electronic switches must be sufficiently low compared to the resistor value R to prevent it from having too much influence on the output signal.

Actual D/A converters are in fact a combination of a ladder network type converter with constant current sources added for the upper and lower bits.

Single bit D/A converter

The disadvantages of the weighted current D/A converter can be overcome by using advanced laser trimming techniques to accurately match resistors

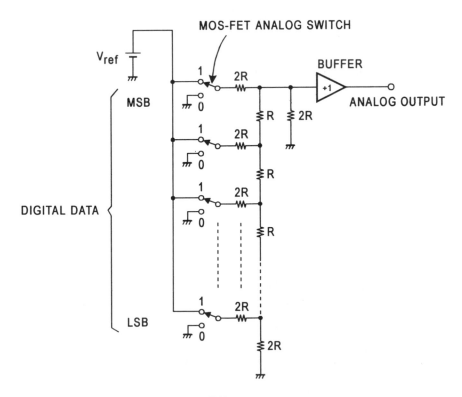

Figure 5.14 *Ladder network type D/A converter*

and current sources. However, this has a negative influence on manufacturing costs. The single bit converter makes high precision D/A conversion possible without the need for expensive matched components.

Figure 5.15 shows the block diagram of the single bit D/A converter. The first step in the conversion process is the requantization of the multi-bit digital audio signal at F_s1 to 1-bit signal with a much higher sampling frequency F_s2. Operation of the requantizer is explained in Fig. 5.16. The requantizer will determine whether the input data D_i is higher or lower than the reference data (D_r), the reference data being the requantized result of the previous input data. Whenever the reference data is lower than the input data, the 1-bit output of the requantizer will be 1 and the reference data will be incremented. This process is repeated for every block cycle of the sam-

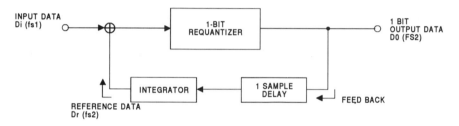

Figure 5.15 *Single bit D/A converter*

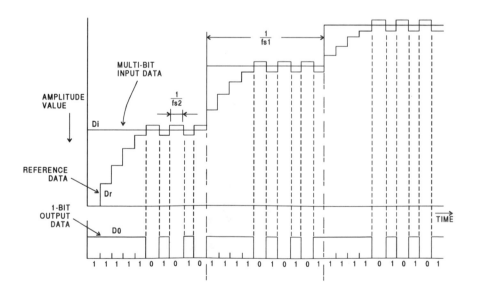

Figure 5.16 *Operation of the 1-bit requantizer*

pling frequency F_s2. At a certain moment the reference data will be higher than the input data, the resulting 1-bit data now becomes 0 and the reference data decremented. The output sequence 0–1–0 continues until new input data is applied to the converter.

The 1-bit output is converted to a pulse width modulation (PWM) or pulse density modulation (PDM) signal which, after low-pass filtering, represents the analog output voltage (Fig. 5.17). Because of feedback of the reference data the single bit D/A converter acts as a noise shaper,

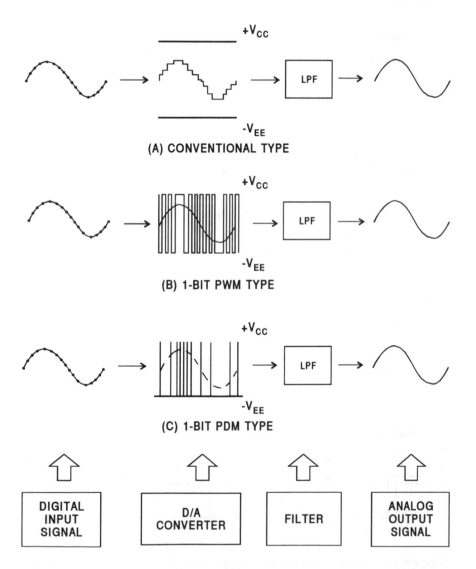

Figure 5.17 *Output signal of (a) conventional D/A converter, (b) 1-bit PWM converter and (c) 1-bit PDM converter*

similar to the single bit A/D converter (Fig. 5.11). It shifts the requantization noise to the higher frequency regions, thereby lowering the quantization
noise in the audible frequency range. Hence, the 1-bit D/A converter easily
achieves a resolution of more than 16-bits in the audio frequency range.
Non-linear distortion and DC offset are non-existing problems. The only
major disadvantage being high frequency noise radiation caused by the high
sampling frequency F_s2, which is usually several MHz. A carefully designed
single bit D/A converter therefore is an inexpensive alternative for high
precision D/A conversion.

Oversampling

The output of a digital-to-analog converter cannot be used directly; filtering
is necessary. The converter output produces the frequency spectrum shown
in Figure 5.18, where the baseband audio signal $(0-f_m)$ is reproduced
symmetrical around the sampling frequency (f_s) and its harmonics. The
low-pass reconstruction filter must reject everything except the baseband
signal.

A sampling frequency (f_s) of 44,100 Hz and a maximum audio frequency
(f_m) of 20,000 Hz mean that a low-pass filter with a flat response to 20 kHz and
a high attenuation at $f_s - f_m$ $(44,100 - 20,000 = 24,100 \, Hz)$ is needed. An
analog filter can be made to have such a sharp roll-off, but the phase response
will introduce an audible phase distortion and group delay.

One approach to getting round this problem is **oversampling**.

Oversampling is the use of a sampling rate greatly in excess of that stipu-
lated by the Nyquist theorem. Practical implementations use a ×2 oversampling $(f_s = 88.2 \, kHz)$ or a ×4 oversampling frequency $(f_s = 176.5 \, kHz)$.
Output spectrum of the D/A converter in a ×2 oversampling system is shown
in figure 5.19, where the large separation between baseband and sidebands
allows a low-pass filter with a gentle roll-off to be used. This improves the
phase response of the filter.

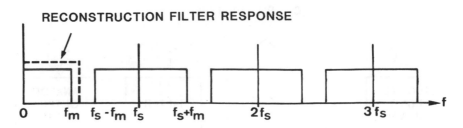

Figure 5.18 *Spectrum of a sample baseband audio signal. A filter must
reject all frequencies above a cut-off frequency f_m*

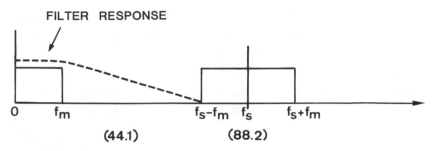

Figure 5.19 *In a ×2 oversampling system the effective sampling frequency becomes twice that of the actual sampling frequency. A simple low-pass filter can be used to reject all unwanted signal frequencies*

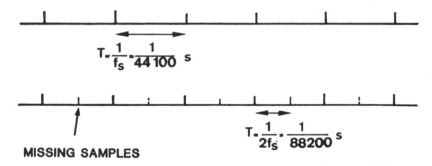

Figure 5.20 *Timing diagram of an oversampling system. Words at a sampling frequency of 44.1 kHz have interpolated samples added, such that the effective sampling rate is 88.2 kHz*

Digital words are input at the standard sampling rate of 44.1 kHz (i.e., no extra samples need be taken at the A/D conversion stage), and extra samples are generated at a rate of 88.2 kHz (Figure 5.20). The missing samples are computed by digital simulation of the analog reconstruction process. A **digital transversal filter** (also known as a **finite impulse response filter**) is well suited for this purpose.

Analog versus digital filters

The discrete-time signal produced by sampling an analog input signal (Figure 5.21) is defined as an infinite series of numbers, each corresponding to a sampling point at time $t = T_n$ for $-\infty < n < +\infty$. Such a series is always referred to by its value at $t = T_n$ which is $x(n)$.

The series $x(n)$ is defined as:

$$x(n) = \ldots, x(-2), x(-1), x(0), x(1), x(2), \ldots$$

with element $x(n)$ occurring at time $t = T_n$.

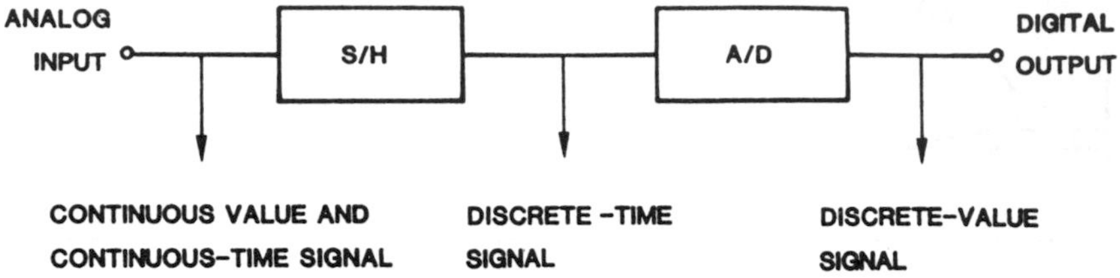

ANALOG INPUT ○——[S/H]——[A/D]——○ DIGITAL OUTPUT

CONTINUOUS VALUE AND CONTINUOUS–TIME SIGNAL

DISCRETE –TIME SIGNAL

DISCRETE–VALUE SIGNAL

Figure 5.21 *Showing how a continuous value and continuous time analog signal is first converted to a discrete time but continuous value set of signals*

Analog filters
The first-order low-pass analog filter shown in figure 5.22 is often described as a function of s, the independent variable in the complex frequency domain. The transfer function of such a filter is given by:

$$f(s) = \frac{V_o}{V_i} = \frac{1}{1+s} = \frac{1}{1+j\dfrac{\omega}{\omega_0}}$$

where:

$$\omega = \text{angular frequency} = 2\pi f$$

and:

ω_0 is the angular frequency at the filter's cut-off frequency $f_c = \dfrac{1}{RC}$

Knowing this, the cut-off frequency of the filter can be calculated as follows:

$$2\pi f_c = \frac{1}{RC}$$

so:

$$f_c = \frac{1}{2\pi RC}$$

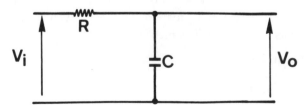

Figure 5.22 *Simple first-order low-pass filter*

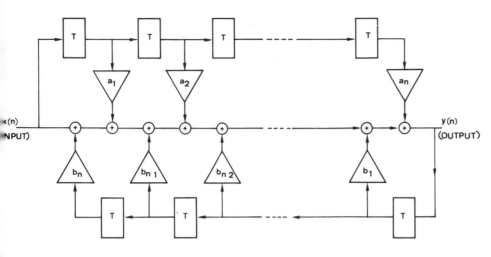

= Unit delay (sampling period)
= Adder
= Multiplier   (coefficient = a_i or $-b_j$)

Figure 5.23 *Recursive digital filter*

Digital filter

A digital filter is a processing system which generates the output sequence, $y(n)$, from an input sequence, $x(n)$, where:

$$y(n) = \underbrace{\sum_{i=0}^{M} a_i\, x(n-i)}_{\substack{\text{present and} \\ \text{past input} \\ \text{samples}}} - \underbrace{\sum_{j=0}^{N} b_j\, y(n-j)}_{\substack{\text{past output} \\ \text{samples}}}$$

The coefficients $a_0, a_1, \ldots, a_M$ and $b_0, b_1, \ldots, b_N$ are constants which describe the filter response.

When $N > 0$, indicating that past output samples are used in the calculation of the present output sample, the filter is said to be **recursive** or **cyclic**. An example is shown in figure 5.23. When only present and past input samples are used in the calculation of the present output sample, the filter is said to be **non-recursive** or **non-cyclic**: because no past output samples are involved in the calculation, the second term then becomes zero (as $N = 0$). An example is shown in figure 5.24

Generally, digital audio systems use non-recursive filters and an example, used in the CDP-102 compact disc player, is shown in figure 5.25 as a block diagram. IC309 is a CX23034, a 96th-order filter which contains 96 multipliers. The constant coefficients are contained in a ROM look-up table. Also note that the CX23034 operates on 16-bit wide data words, which means that all adders and multipliers are 16-bit devices.

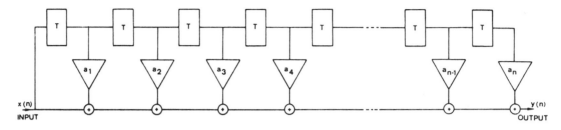

Figure 5.24 *Non-recursive digital filter*

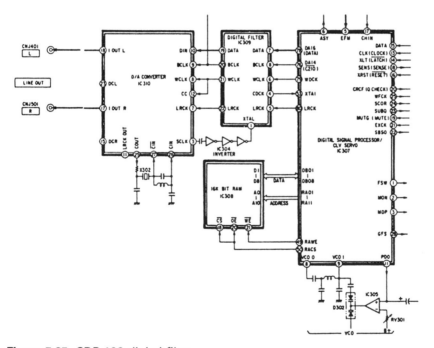

Figure 5.25 *CDP-102 digital filter*

6
Codes for Digital Magnetic Recording

The binary data representing an audio signal can be recorded on tape (or disc) in two ways: either directly, or after frequency modulation.

When frequency modulation is used, say, in helical-scan recorders, data can be modulated as they are, usually in a **non-return-to-zero** format (see further). If they are recorded directly, however, say, in stationary head recorders and compact disc, they have to be transformed to some new code to obtain a recording signal which matches as well as possible the properties of the recording channel.

This code should have a format which allows the highest bit density permitted by the limiting characteristics of the recording channel (frequency response, dropout rate, etc.) to be obtained. Also, its DC content should be eliminated, as magnetic recorders cannot reproduce DC.

Coding of binary data in order to comply with the demands of the recording channel is often referred to as **channel coding**.

Non-Return to Zero (NRZ)

This code is one of the oldest and best known of all channel codes. Basically, a logic 1 is represented by positive magnetization, and a logic 0 by negative magnetization.

A succession of the same logic levels, though, presents no change in the signal, so that there may be a significant low-frequency content, which is undesirable for stationary-head recording.

In helical-scan recording techniques, on the other hand, the data are FM-converted before being recorded, so this property is less important. NRZ is commonly used in such formats as PCM-1600 and the EIAJ-format PCM-100 and PCM-10 recorders.

Several variations of NRZ also exist for various applications.

Bi-Phase

Similar to NRZ, but extra transitions are added at the beginning of every data bit interval. As a result, DC content is eliminated and synchronization becomes easier, but the density of signal transitions increases.

This code (and its variants) is also known as **Manchester code**, and is used in the video 8 PCM recording format, where bits are modulated as a 2.9 MHz signal for a logic 0 and as 5.8 MHz for a logic 1.

Modified Frequency Modulation (MFM)

Also called **Miller code** or **delay modulation**. Ones are coded with transitions in the middle of the bit cell, isolated zeros are ignored, and between pairs of zeros a transition is inserted. It requires almost the same low bandwidth as NRZ, but has a reduced DC content. The logic needed for decoding is more complicated.

A variation is the so-called **modified modified frequency modulation** (M^2FM).

3-Position Modulation (3PM)

This is a code which permits very high packing densities, but which requires rather complicated hardware. In principle, 3PM code is obtained by dividing the original NRZ data into blocks of 3; each block is then converted to a 6-bit 3PM code, which is designed to optimize the maximum and minimum run lengths. In this way, the minimum possible time between transitions is two times the original (NRZ) clock period, whereas the maximum is six times the original.

For detection, on the other hand, a clock frequency twice that of the original signal is needed, consequently reducing the jitter margin of the system. This clock is normally recovered from the data itself, which have a high harmonic content around the clock frequency.

High Density Modulation – 1 (HDM – 1)

This is a variation upon the 3PM system. The density ratio is the same as 3PM, but clock recovery is easier and the required hardware simpler. It is proposed by Sony for stationary-head recording.

Eight-to-Fourteen Modulation (EFM)

This code is used for the compact disc digital audio system. The principle is similar again to 3PM, but each block of 8 data bits is converted into 14 channel bits, to which 3 extra bits are added for merging (synchronization) and low-frequency suppression. In this way, a good compromise is obtained between clock accuracy (and possible detection errors), minimum DC current (in disc systems, low frequencies in the signal give noise in the servo systems), and hardware complexity. Also this modulation system is very suitable for combination with the error-correction system used in the same format.

7
Principles of Error Correction

Types of Code Errors

When digital signals are sent through a transmission channel, or recorded and subsequently played back, many types of code errors can occur. Now, in the digital field, even small errors can cause disastrous audible effects: even one incorrectly detected bit of a data word will create an audible error if it is, say, the MSB of the word. Such results are entirely unacceptable in the high quality expected from digital audio, and a lot of effort must be made to detect, and subsequently correct, as many errors as possible without making the circuit over-complicated or the recorded bandwidth too high.

There are a number of causes of code errors:

Dropouts

Dropouts are caused by dust or scratches on the magnetic tape or CD surface, or microscopic bubbles in the disc coating.

Tape dropout causes relatively long-time errors, called **bursts**, in which long sequences of related data are lost together. On discs dropouts may cause either burst or single random errors.

Jitter

Tape jitter causes random errors in the timing of detected bits and, to some extent, is unavoidable, due to properties of the tape transportation mechanism. **Jitter margin** is the maximum amount of jitter which permits correct detection of data. If the minimum run length of the signal is τ, then the jitter margin will be $\tau/2$. Figure 7.1 shows this with an NRZ signal.

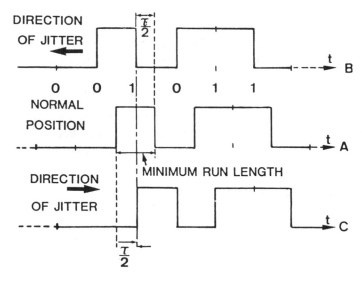

Figure 7.1 *Illustrating jitter margin*

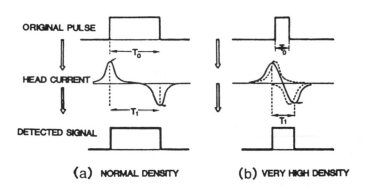

Figure 7.2 *Illustrating the cause of intersymbol interference*

Intersymbol interference

In stationary-head recording techniques, a pulse is recorded as a positive current followed by a negative current (see Figure 7.2). This causes the actual period of the signal that is read on the tape (T_1) to be longer than the bit period itself (T_0). Consequently, if the bit rate is very high, the detected pulse will be wider than the original pulse. Interference, known as intersymbol **interference** or **time crosstalk** may occur between adjacent bits. Intersymbol interference causes random errors, depending upon the bit situation.

Noise

Noise may have similar effects to dropouts (differentiation between both is often difficult), but random errors may also occur in the case of pulse noise.

Editing

Tape editing always destroys some information on the tape, which consequently must be corrected. Electronic editing can keep errors to a minimum, but tape-cut editing will always cause very long and serious errors.

Error Compensation

Errors must be detected by some error-detection mechanism: if misdetection occurs, the result is audible disturbance. After detection, an error-correction circuit attempts to recover the correct data. If this fails, an error-concealment mechanism will cover up the faulty information so, at least, there is no audible disturbance.

These three basic functions: detection, correction and concealment, are illustrated in Figure 7.3.

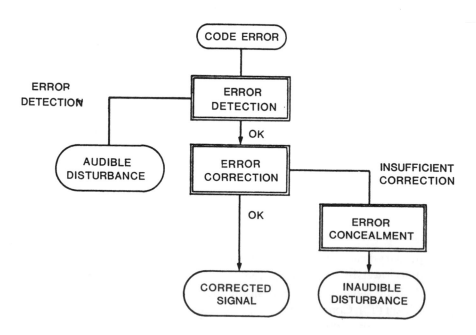

Figure 7.3 *Three basic functions of error compensation: detection, correction and concealment*

Error Detection

Simple parity checking

To detect whether a detected data word contains an error, a very simple way is to add one extra bit to it before transmission. The extra bit is given the value 0 or 1, depending upon the number of 1s in the word itself. Two systems are possible: **odd parity** and **even parity** checking.

- Odd parity: where the total number of 1s is odd.
  Example:

  |1110| |0|
  data   parity

  |1001| |1|
  data   parity

- Even parity: where the total number of 1s is even.
  Example:

  |1110| |1|
  data   parity

  |1001| |0|
  data   parity

The detected word must also have the required number of 1s. If this is not the case, there has been a transmission error.

This rather elementary system has two main disadvantages:

1 even if an error is detected, there is no way of knowing which bit was faulty
2 if two bits of the same word are faulty, the errors compensate each other and no errors are detected.

Extended parity checking

To increase the probability of detecting errors, we can add more than one parity bit to each block of data. Figure 7.4 shows a system of extended parity, in which M-1 blocks of data, each of n-bits, are followed by block m — the parity block. Each bit in the parity block corresponds to the relevant bits in each data block, and follows the odd or even parity rules outlined previously.

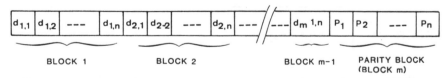

Figure 7.4 *Extended parity checking*

If the number of parity bits is n, it can be shown that (for reasonably high values of n) the probability of detecting errors is $\frac{1}{2}^n$.

Cyclic redundancy check code (CRCC)

The most efficient error-detection coding system used in digital audio is **cyclic redundancy check code** (CRCC), which relies on the fact that a bit stream of n bits can be considered an algebraic polynomial in a variable x with n terms. For example, the word 10011011 may be written as follows:

$$M(x) = 1x^7 + 0x^6 + 0x^5 + 1x^4 + 1x^3 + 0x^2 + 1x^1 + 1x^0$$
$$= x^7 + x^4 + x^3 + x + 1$$

Now to compute the cyclic check on $M(x)$, another polynomial $G(x)$ is chosen. Then, in the CRCC encoder, $M(x)$ and $G(x)$ are divided:

$$M(x)/G(x) = Q(x) + R(x)$$

where $Q(x)$ is the quotient of the division, and $R(x)$ is the remainder.
Then (Figure 7.5), a new message $U(x)$ is generated as follows:

$$U(x) = M(x) + R(x)$$

so that $U(x)$ can always be divided by $G(x)$ to produce a quotient with no remainder.

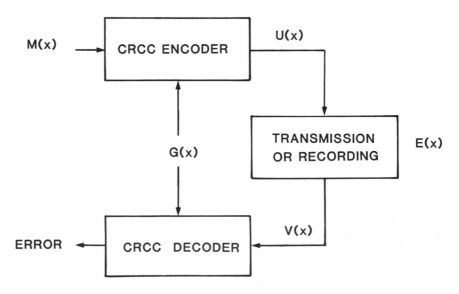

Figure 7.5 *CRCC checking principle*

It is this message $U(x)$ that is recorded or transmitted. If, on playback or at the receiving end, an error $E(x)$ occurs, the message $V(x)$ is detected instead of $U(x)$, where:

$$V(x) = U(x) + E(x)$$

In the CRCC decoder, $V(x)$ is divided by $G(x)$, and the resultant remainder $E(x)$ shows there has been an error.

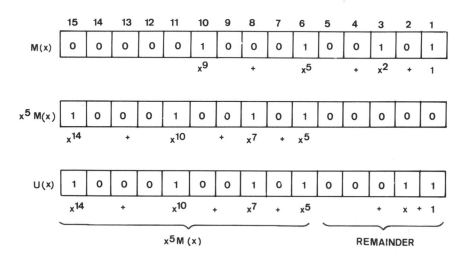

Figure 7.6 *Generation of a transmission polynomial*

Example (illustrated in Figure 7.6):

the message is $M(x) = x^9 + x^5 + x^2 + 1$
the check polynomial is $G(x) = x^5 + x^4 + x^2 + 1$

Now, before dividing by $G(x)$ we multiply $M(x)$ by x^5; or, in other words, we shift $M(x)$ five places to the left, in preparation of the five check bits that will be added to the message:

$$x^5 M(x) = x^{14} + x^{10} + x^7 + x^5$$

Then the division is made:

$$x^5 M(x)/G(x) = \frac{(x^9 + x^8 + x^7 + x^3 + x^2 + x + 1)}{\text{quotient}} + \frac{(x + 1)}{\text{remainder}}$$

So that:

$$U(x) = x^5 M(x) + (x + 1)$$
$$= x^{14} + x^{10} + x^7 + x^5 + x + 1$$

which can be divided by $G(x)$ to leave no remainder.

Figure 7.6 shows that, in fact, the original data are unmodified (only shifted), and that the check bits follow at the end.

CRCC checking is very effective in detection of transmission error. If the number of CRCC bits is n, detection probability is $1 - 2^{-n}$. If, say, n is 16, as in the case of the Sony PCM-1600, detection probability is $1 - 2^{-16} =$ 0.999985 or 99.9985%. This means that the CRCC features almost perfect detection capability. Only if $E(x)$, by coincidence, is exactly dividable by $G(x)$, will no error be detected. This obviously occurs only rarely and, knowing the characteristics of the transmission (or storage) medium, polynomial $G(x)$ can be chosen such that the possibility is further minimized.

Although CRCC error-checking seems rather complex, the divisions can be done relatively simply using modulo-2 arithmetic. In practical systems LSIs are used which perform the CRCC operations reliably and fast.

Error Correction

In order to ensure later correction of binary data, the input signal must be encoded by some encoding scheme. The data sequence is divided into message blocks, then, each message block is transformed into a longer one, by adding additional information, in a process called **redundancy**.

The ratio $\dfrac{\text{data} + \text{redundant data}}{\text{data}}$ is known as the **code rate**.

There is a general theory, called the **coding theorem**, which says that the probability of decoding an error can be made as small as possible by increasing the code length and keeping the code rate less than the channel capacity.

When errors are to be corrected, we must not only know that an error has occurred, but also exactly which bit or bits are wrong. As there are only two possible bit states (0 or 1), correction is then just a matter of reversing the state of the erroneous bits.

Basically, correction (and detection) of code errors can be achieved by adding to the data bits an additional number of redundant check bits. This redundant information has a certain connection with the actual data, so that, when errors occur, the data can be reconstructed again. The redundant information is known as the error-correction code. As a simple example, all data could be transmitted twice, which would give a redundancy of 100%. By comparing both versions, or by CRCC, errors could easily be detected, and if some word were erroneous, its counterpart could be used to give the correct data. It is even possible to record everything *three* times; this would be still more secure. These are however rather wasteful systems, and much more efficient error-correction systems can be constructed.

The development of strong and efficient error-correction codes has been one of the key research points of digital audio technology. A lot of experience has been used from computer technology, where the correction of code errors is equally important, and where a lot of research has been spent in order to optimize correction capabilities of error-correction codes. Design of strong codes is a very complex matter however, which requires thorough study and use of higher mathematics: algebraic structure has been the basis of the most important codes. Some codes are very strong against 'burst' errors, i.e., when entire clusters of bits are erroneous together (such as during tape dropouts), whereas others are better against 'random' errors, i.e., when single bits are faulty.

Error-correction codes are of two forms, in which:

1 data bits and error-correction bits are arranged in separate blocks; in this case we talk about **block codes**. Redundancy that follows a data block is only generated by the data in that particular block
2 data and error correction are mixed in one continuous data stream; in this case we talk about **convolutional codes**. Redundancy within a certain time unit does not only depend upon the data in that same time unit, but also upon data occurring a certain time before. They are more complicated, and often superior in performance to block codes.

Figure 7.7 illustrates the main differences between block and convolutional error-correcting codes.

DATA 1	REDUNDANCY 1	DATA 2	REDUNDANCY 2

(a) Block Code

DATA   REDUNDANCY

(b) Convolutional Code

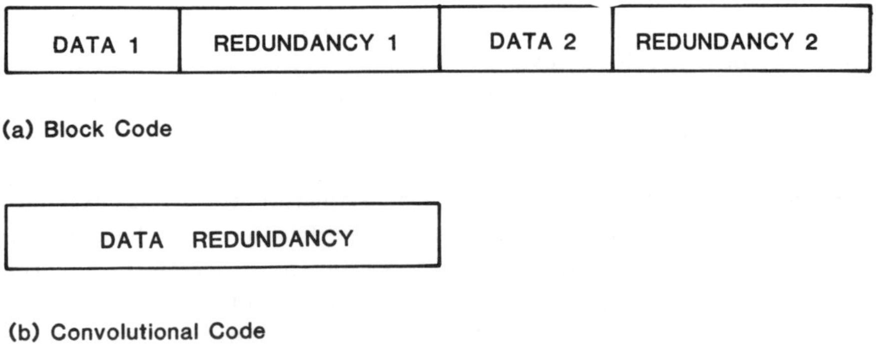

Figure 7.7 *Main differences between block and convolutional error-correcting codes*

Combinational (horizontal/vertical) parity checking

If, for example, we consider a binary word or message consisting of 12 bits, these bits could be arranged in a 3×4 matrix as shown in Figure 7.8.

Then, to each row and column one more bit can be added to make parity for that row or column even (or odd). Then, in the lower right-hand corner, a final bit can be added that will give the last column an even parity as well; because of this the last row will also have even parity.

$$
\begin{array}{cccc|c}
1 & 0 & 1 & 0 & 0 \\
1 & 0 & 1 & 1 & 1 \\
0 & 0 & 1 & 0 & 1 \\
\hline
0 & 0 & 1 & 1 & 0 \\
\end{array}
$$

Figure 7.8 *Combinational parity checking*

If this entire array is transmitted in sequence (row by row, or column by column), and if during transmission an error occurs to one bit, parity check on one row and on one column will fail; the error will be found at the intersection and, consequently, it can be corrected. The entire array of 20 bits, of which 12 are data bits, form a code word, which is referred to as a (20,12) code. There are $20 - 12 = 8$ redundant digits.

All error-correcting codes are more or less based on this idea, although better codes, i.e., codes using fewer redundant bits than the one given in our example, can be constructed.

Crossword code

Correction of errors to single bits is just one aspect of error-correcting codes. In digital recording, very often errors come in bursts, with complete clusters of faulty bits. It will be obvious that, in view of the many possible combinations, the correction of such bursts or errors is very complicated and demands powerful error-correcting codes.

One such code, developed by Sony for use in its PCM-1600 series, is the **crossword code**. This uses a matrix scheme similar to the previous example, but it carries out its operations with whole words instead of individual bits, with the resultant advantage that large block code constructions can be easily realized so that burst error correction is very good. Random error correction, too, is good. Basically, the code allows detection and correction of all types of errors with a high probability of occurrence, and that only errors with a low probability of occurrence may pass undetected.

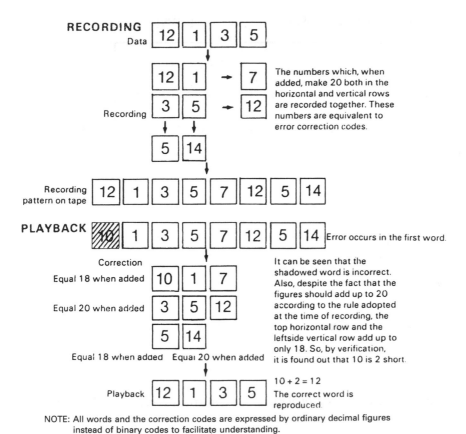

Figure 7.9 *Illustration of a crossword code*

A simple illustration of a crossword code is given in Figure 7.9, where the decimal numbers represent binary values or words.

Figure 7.10 shows another simple example of the crossword code, in binary form, in which four words M_1 to M_4 are complemented in the coder by four parity or information words R5 to R8, so that:

$$R_5 = M_1 \oplus M_2$$
$$R_6 = M_3 \oplus M_4$$
$$R_7 = M_1 \oplus M_3$$
$$R_8 = M_2 \oplus M_4$$

where the symbol $\oplus$ denotes modulo-2 addition.

All eight words are then recorded, and at playback received as U_1 to U_8.

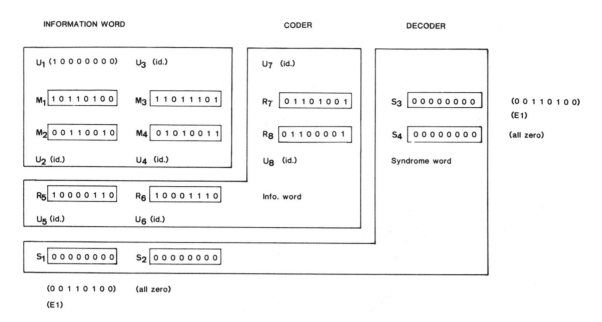

Figure 7.10 *Crossword code using binary data*

In the decoder, additional words are constructed, called syndrome words, as follows:

$$S_1 = U_1 \oplus U_2 \oplus U_5$$
$$S_2 = U_3 \oplus U_4 \oplus U_6$$
$$S_3 = U_1 \oplus U_3 \oplus U_7$$
$$S_4 = U_2 \oplus U_4 \oplus U_8$$

By virtue of this procedure, if all received words U_1 to U_8 are correct, all syndromes must be zero. On the other hand, if an error E occurs in one or more words, we can say that:

$$U_i = M_i \oplus E_i \text{ for } i = 1 \text{ to } 4$$
$$U_i = R_i \oplus E_i \text{ for } i = 5 \text{ to } 8$$

Now:

$$S_1 = U_1 \oplus U_2 \oplus U_5$$
$$= M_1 \oplus E_1 \oplus M_2 \oplus E_2 \oplus R_5 \oplus E_5$$
$$= E_1 \oplus E_2 \oplus E_5$$

as we know that $M_1 \oplus M_2 \oplus R_5 = 0$. Similarly:

$S_2 = E_3 \oplus E_4 \oplus E_6$
$S_3 = E_1 \oplus E_3 \oplus E_7$
$S_4 = E_2 \oplus E_4 \oplus E_8$

Correction can then be made, because:

$U_1 \oplus S_3 = M_1 \oplus E_1 \oplus E_1 = M_1$

Of course, there is still a possibility that simultaneous errors in all words compensate each other to give the same syndrome patterns as in our example. The probability of this occurring, however, is extremely low and can be disregarded.

In practical decoders, when errors occur, the syndromes are investigated and a decision is made whether there is a good probability of successful correction. If the answer is yes, correction is carried out; if the answer is no, a concealment method is used. The algorithm the decision-making process follows must be initially decided using probability calculations, but once it is fixed it can easily be implemented, for instance, in a P-ROM.

Figure 7.11 shows the decoding algorithm for this example crossword code. Depending upon the value of the syndrome(s), decisions are made for correction according to the probability of miscorrection; the right column shows the probability of each situation occurring.

b-adjacent code

A code which is very useful for correcting both random and burst errors has been described by D. C. Bossen of IBM, and called **b-adjacent code**. The b-adjacent error-correction system is used in the EIAJ format for home-use helical-scan digital audio recorders.

In this format two parity words, called P and Q, are constructed as follows:

$P_n = L_n \oplus R_n \oplus L_{n+1} \oplus R_{n+1} \oplus R_{n+2}$
$Q_n = T^6 L_n \oplus T^5 R_n \oplus T^4 L_{n+1} \oplus T^3 R_{n+1} \oplus T^2 L_{n+2} \oplus T R_{n+2}$

where T is a specific matrix of 14 words of 14 bits; L_n, R_n, etc. are data words from, respectively, the left and the right channel.

CIRC code and other codes

Many other error-correcting codes exist, as most manufacturers of professional audio equipment design their own preferred error-correction system.

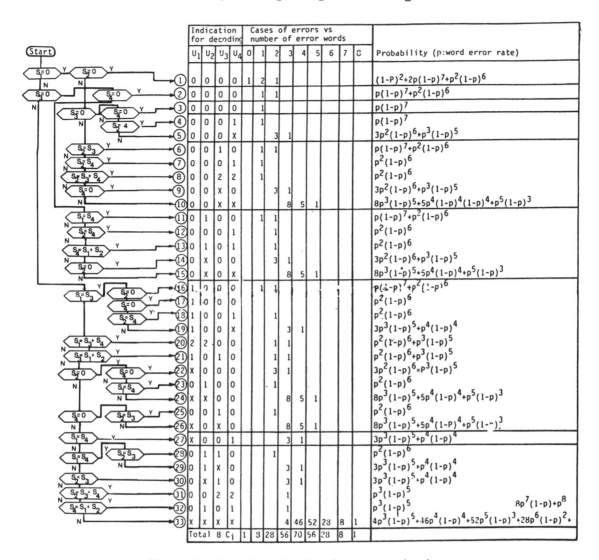

Figure 7.11 *Decoding algorithm for crossword code*

Most, however, are variations on the best-known codes, with names such as Reed-Solomon code, BCH (Bose-Chaudhuri-Hocquenghem) code after the researchers who invented them.

Sony developed (together with Philips) the CIRC (cross-interleave Reed-Solomon code) for the compact disc system. R-DAT tape format uses a double Reed-Solomon code, for extremely high detection and correction capability. The DASH format for professional stationary head recorders uses a powerful combination of correction strategies, to allow for quick in/out and editing (both electric and tape splice).

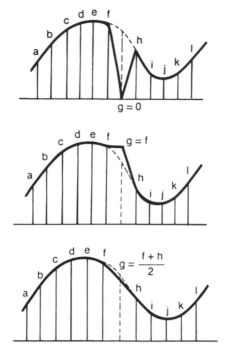

Figure 7.12 *Three types of error concealment (a) muting (b) previous word holding (c) linear interpolation*

Error Concealment

Next comes a technique which prevents the uncorrected code errors from affecting the quality of the reproduced sound. This is known as **concealment**, and there are 4 typical methods.

1 **muting:** the erroneous word is muted, i.e., set to zero (Figure 7.12a). This is a rather rough concealment method, and consequently used rarely

2 **previous word holding:** the value of the word before the erroneous word is used instead (Figure 7.12b) so that there is usually no audible difference. However, especially at high frequencies where sampling frequency is only 2 or 3 times the signal frequency, this method may give unsatisfactory results

3 **linear interpolation** (also called **averaging**): the erroneous word is replaced by the average value of the preceding and succeeding words (Figure 7.12c). This method gives much better results than the two previous methods

4 **higher-order polynomial interpolation:** gives an estimation of the correct value of the erroneous word by taking into account a number of preceding and following words. Although much more complicated than previous methods, it may be worthwhile to use it in very critical applications.

Interleaving

In view of the high recording density used to record digital audio on magnetic tape, dropouts could destroy many words simultaneously, as a burst error. Error correction that could cope with such a situation would be prohibitively complicated and, if it failed, concealment would not be possible as methods like interpolation demand that only one sample of a series is wrong.

For this reason, adjacent words from the A/D converter are not written next to each other on the tape, but at locations a sufficient distance apart to make sure that they cannot suffer from the same dropout. In effect, this method converts possible long burst errors into a series of random errors. Figure 7.13 illustrates this in a simplified example. Words are arranged in **interleave blocks**, which must be at least as long as the maximum burst error. Practical interleaving methods are much more complicated than this example, however.

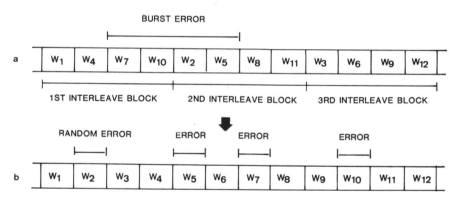

Figure 7.13 *Showing how interleaving of data effectively changes burst errors into random errors*

PART TWO

The Compact Disc

8
Overview of the Compact Disc Medium

It is the compact disc that has introduced most people to digital audio reproduction. Table 8.1 is a comparison of LP and CD systems, showing that CD is far superior to LP in each aspect of dynamic range, distortion, frequency response, and wow and flutter specifications. In particular, CD exhibits a remarkably wide dynamic range (90 dB) throughout the entire audible frequency spectrum. In contrast, dynamic range of LP is 70 dB at best. Harmonic distortion of CD reproduction is less than 0.01%, which is less than one hundredth of that of LP. Wow and flutter are simply too minute to be measured in a CD system. This is because, in playback, digital data are first stored in a RAM and then released in perfect, uniform sequence determined by a reference clock of quartz precision.

With a mechanical system like that of LP, the stylus must be in physical contact with the disc. Therefore, both the stylus and the disc will eventually wear out, causing serious deterioration of sound quality. With the CD's optical system, however, lack of contact between the disc and the pick-up means that there is no sonic deterioration no matter how many times the disc is played.

Mechanical (and, for that matter, variable capacitance) systems are easily affected by dust and scratches, as signals are impressed directly on the disc surface. A compact disc, however, is covered with a protective layer (the laser optical pickup is focused underneath this) so that the effect of dust and scratches is minimized. Furthermore, a powerful error-correction system, which can correct even large burst errors, makes the effect of even severe disc damage insignificant in practice.

Table 8.1 *System comparison between CD and LP*

	CD system	Conventional LP player
Specifications		
Frequency response	20 Hz–20 kHz ± 0.5 dB	30 Hz–20 kHz ± 3 dB
Dynamic range	More than 90 dB	70 dB (at 1 kHz)
S/N	90 dB (with MSB)	60 dB
Harmonic distortion	Less than 0.01%	1–2%
Separation	More than 90 dB	25–30 dB
Wow and flutter	Quartz precision	0.03%
Dimensions		
Disc	12 cm (diameter)	30 cm (diameter)
Playing time (on one side)	60 minutes (maximum 74 minutes)	20–25 minutes
Operation/reliability		
Durability disc	Semi-permanent	High-frequency response is degraded after being played several tens of times
Durability stylus	Over 5000 hours	500–600 hours
Operation	– Quick and easy access due to micro computer control – A variety of programmed play possible – Increased resistivity to external vibration	– Needs stylus pressure adjustment – Easily affected by external vibration
Maintenance	Dust, scratches, and fingerprints are made almost insignificant	Dust and scratches cause noise

Photo 8.1 *CD player*

Main Parameters

Main parameters of CD compared to LP are shown in Table 8.2. Figure 8.1 compares CD and LP disc sizes. Figure 8.2 compares track pitch and groove dimensions of a CD with an LP; 60 tracks of the CD would fit into one track of an LP.

Table 8.2 *Parameter comparison between CD and LP*

	CD	LP
Disc diameter	120 mm	305 mm
Rotation speed	568–228 rpm (at 1.4 ms^{-1}) 486–196 rpm (at 1.2 ms^{-1})	33⅓ rpm
Playing time (maximum)	74 min	32 min (one side)
No. of tracks	20,625	1060 maximum
Track spacing	1.6 μm	85 μm
Lead-in diameter	46 mm	302 mm
Lead-out diameter	116 mm	121 mm
Total track length	5300 mm	705 m maximum
Linear velocity	1.2 or 1.4 ms^{-1}	528–211 mms^{-1}

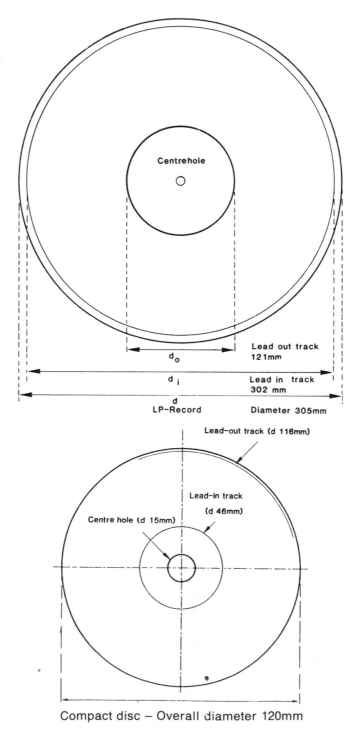

Figure 8.1 *A comparison of CD and LP sizes*

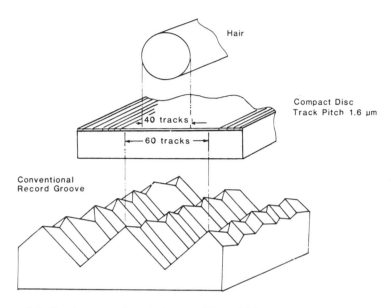

Figure 8.2 *Track comparison between CD and LP*

Optical Discs

Figure 8.3 gives an overview of the optical discs available. The CD single is the digital equivalent of a 45 rpm single. It can contain about 20 minutes of music and is fully compatible with any CD player. A CD Video (CDV) contains 20 minutes of digital audio which can be played back on an ordinary CD player and 6 minutes of video with digital audio. To play back the video part you need a Video Disc Player or a Multi Disc Player (MDP).

Multi Disc Players are capable of playing both Compact Discs and Laser

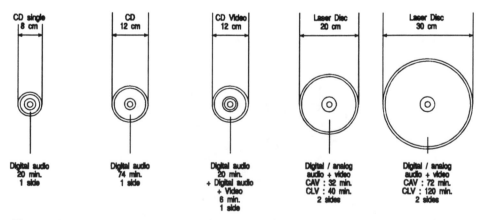

Figure 8.3 *Optical discs*

Discs (LD). Optical discs containing video signals can be distinguished from disc containing only digital audio by their colour. CDV and LD have a gold shine, while CD and CD single have a silver shine.

Recording and Readout System on a CD

The data on a compact disc are recorded on the master by using a laser beam photographically to produce pits in the disc surface, in a clockwise spiral track starting at the centre of the disc. Length of the pits and the distance between them from the recorded information, are as shown in figure 8.4

In fact, on the user disc, the pits are actually bumps. These can be identified by focusing a laser beam onto the disc surface: if there is no bump on the surface, most of the light that falls on the surface (which is highly reflective) will return in the same direction. If there is a bump present however, the light will be scattered and only a small portion will return in the original direction (Figure 8.5). The disc has a 1 mm thick protective transparent layer over the signal layer, i.e., the pits. More important, the spot size of the laser beam is about 1 mm in diameter at the surface of the disc, but is as small as $1.7\,\mu$m across at the signal layer. This means that a dust particle or a scratch on the disc surface is literally out of focus to the sensing mechanism. Figure 8.6 illustrates this. Obviously, control of focus must be extremely accurate.

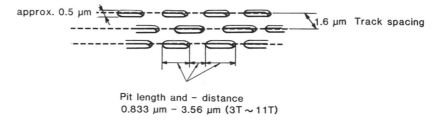

Pit length and – distance
0.833 µm – 3.56 µm (3T ~ 11T)

Figure 8.4 *Pits on a CD, viewed from label side*

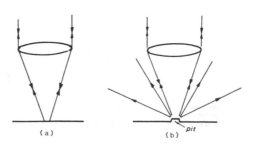

Figure 8.5 *CD laser beam reflection (a) from disc surface (b) from pit*

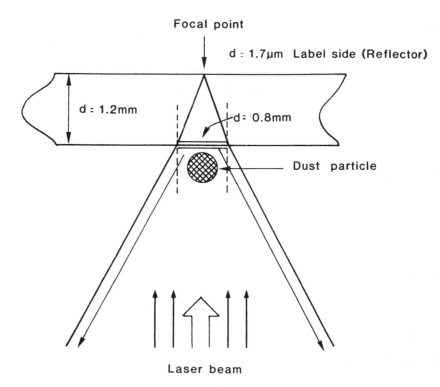

Figure 8.6 *Showing how a dust particle on the disc surface is out of focus*

Signal Parameters

Before being recorded, the digital audio signal (which is a 16-bit signal) must be extended with several additional items of data. These include:

● error correction data
● control data (time, titles, lyrics, graphics and information about the recording format or emphasis)
● synchronization signals, used to detect beginning of each data block
● merging bits: added between each data symbol to reduce the DC component of the output signal.

Audio Signal

The audio signal normally consists of two channels of audio, quantized with a 16-bit linear quantization system at a sampling frequency of 44.1 kHz. During

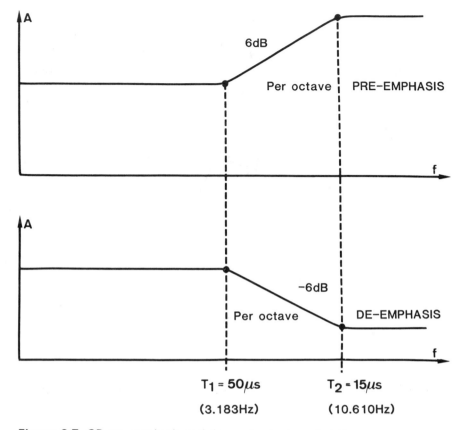

Figure 8.7 *CD pre-emphasis and de-emphasis characteristics*

recording, pre-emphasis (slight boost of the higher frequencies) may be applied. Pre-emphasis standards agreed for the compact disc format are 50 μs and 15 μs (or 3183 Hz and 10,610 Hz).

Consequently, the player must in this case apply a similar de-emphasis to the decoded signal to obtain a flat frequency response (Figure 8.7).

A specific control code recorded along with the audio signal on the compact disc is used to inform the player whether pre-emphasis is used, and so the player switches in the corresponding de-emphasis circuit to suit.

Alternatively, audio information on the CD may comprise four music channels instead of two; this is also identified by a control code to allow automatic switching of players equipped with a 4-channel playback facility. Although, on launching CD, there were no immediate plans for 4-channel discs or players, the possibility for later distribution was already provided in the standard.

Additional information on the CD

Before the start of the music programme, a 'lead-in' signal is recorded on the CD. When a CD is inserted, most players immediately read this lead-in signal which contains a 'table of contents' (TOC). The TOC contains information on the contents of the disc, such as the starting-point of each selection or track, number of selections, duration of each selection. This information can be displayed on the player's control panel, and/or used during programme search operation.

At the end of the programme, a lead-out signal is similarly recorded which informs the player that playback is complete.

Furthermore, **music start flags** between selections inform the player that a new selection follows.

Selections recorded on the disc can be numbered from 1 through 99. In each track, up to 99 indexes can be given, which may separate specific sections of the selection. Playing time is also encoded on the disc in minutes, seconds and 1/75ths of a second; before each selection, this time is counted down.

There is further space available to encode other information, such as: titles, performer names, lyrics and even graphic information which may all be displayed, for instance on a TC screen during playback.

Compact Disc Production

Compact disc 'cutting'

Figure 8.8 is a block diagram comparing CD digital audio recording and playback systems with analog LP systems.

The two systems are quite similar and, in fact, overlap can occur at record production stage. However, where LP masters are mechanically cut, CD masters are 'cut' in an electro-optical photographic process: no 'cutting' of the disc surface actually takes place.

The CD disc production process follows seven main stages, illustrated in Figure 8.9:

1 a glass plate is polished for optimum smoothness
2 a photo-resist coating is applied to its surface. The roughness of the glass surface and the thickness of the coating determines the depth of the pits on the compact disc
3 the photo-resistive coating is then exposed to a laser beam, the intensity of which is modulated with digitized audio information
4 the photo-resist layer is developed and the pits of information are revealed
5 the surface is silvered to protect the pits
6 the surface is plated with nickel to make a metal master

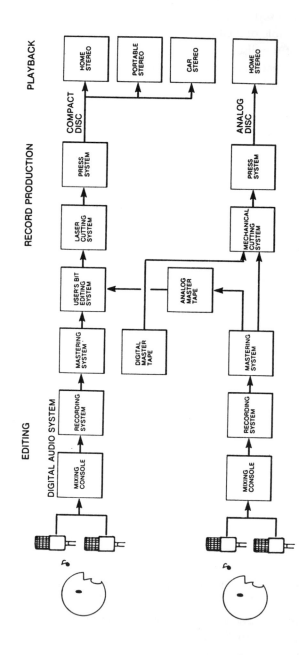

Figure 8.8 *A comparison of CD and LP recording and playback systems*

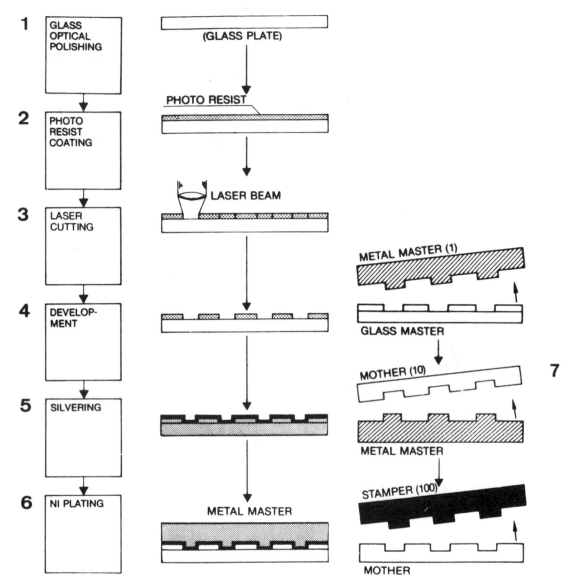

Figure 8.9 *Stages in the 'cutting' of a compact disc*

7 the metal master is then used to make mother plates. These mothers are in turn used to make further metal masters, or stampers.

Compact disc stamping

The stamping process, although named after the analogous stage in LP record production, is, in fact, an injection moulding, compression moulding or polymerization process, producing plastic discs (Figure 8.10). The signal surface of each disc is then coated with a reflective material (vaporized aluminium) to enable optical read-out, and further protected with a transparent plastic layer which also supports the disc label.

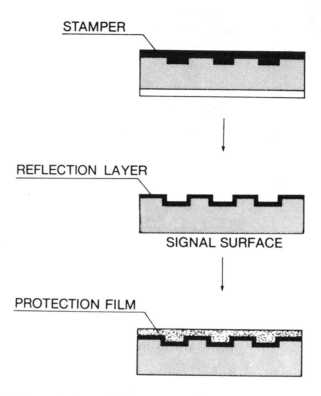

Figure 8.10 *Stages in 'stamping' a compact disc*

9
Compact Disc Encoding

A substantial amount of information is added to the audio data before the compact disc is recorded. Figure 9.1 illustrates the encoding process and shows the various information to be recorded.

There are usually two audio channels with 16-bit coding, sampled at 44.1 kHz. So, the bit rate, after combining both channels, is:

$$44.1 \times 16 \times 2 = 1.4112 \times 10^6 \, \text{bits s}^{-1}$$

CIRC Encoding

Most of the errors which occur on a medium such as CD are random. However, from time to time burst errors may occur due to fingerprints, dust or scratches on the disc surface. To cope with both random and burst errors, Sony and Philips developed the cross interleave Reed-Solomon error-correction code (CIRC). CIRC is a very powerful combination of several error correction techniques.

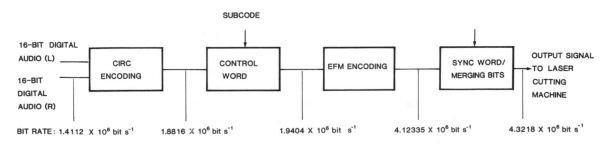

Figure 9.1 *Encoding process in compact disc production*

Table 9.1 *Specifications of CIRC system in the compact disc*

Aspect	Specification
Maximum correctable burst error length	4000 bits (i.e., 2.5 mm on the disc surface)
Maximum interpolatable burst error length	12,300 bits (i.e., 7.7 mm on the disc surface)
Sample interpolation rate	One sample every 10 hours at a BER of 10^{-4}
	1000 samples every minute at a BER of 10^{-3}
Undetected error samples	Less than one every 750 hours at a BER of 10^{-3}
	Negligible at a BER of 10^{-4}
Code rate	On average, four bits are recorded for every three data bits

It is useful to be able to measure an error-correcting system's ability to correct errors, and as far as the compact disc medium is concerned it is the maximum length of a burst error which is critical. Also, the greater the number of errors received, the greater the probability of some errors being uncorrectable. The number of errors received is defined as the **bit error rate** (BER). An important system specification, therefore, is the number of data samples per unit time, called the **sample interpolation rate**, which have to be interpolated (rather than corrected) for given BER values. The lower this rate is, the better the system. Then, if burst errors cannot be corrected, an important specification is the maximum length of a burst error which can be interpolated. Finally, it is important to know the number of undetected errors, resulting in audible clicks. Any specification of an error-correcting system must take all these factors into account. Table 9.1 is a list of all relevant specifications of the CIRC system used in CD.

The CIRC principle is as follows (refer to Figure 9.2):

● the audio signal is sampled (digitized) at the A/D converter and these 16-bit samples are split up into two 8-bit words called **symbols**

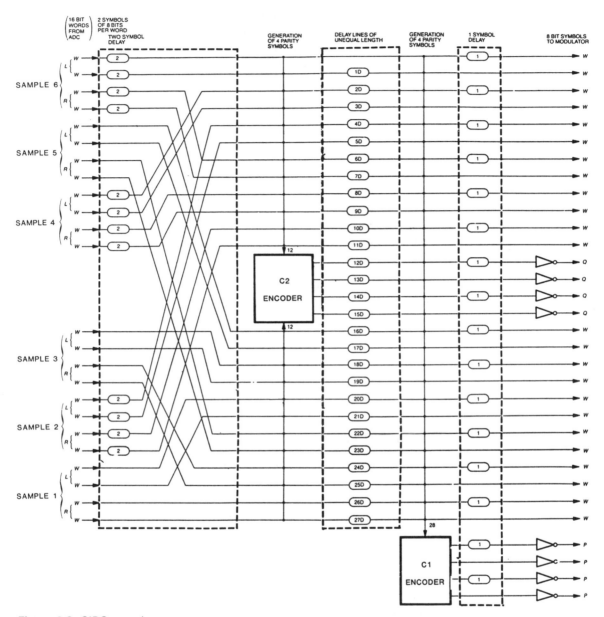

Figure 9.2 *CIRC encoder*

- six of 16-bit samples, from each channel, i.e., twenty-four 8-bit symbols are applied to the CIRC encoder, and stored in a RAM memory
- the first operation in the CIRC encoder is called **scrambling**. The scrambling operation consists of a two-symbol delay for the even samples and a mixing up of the connections to the C2 encoder

- the 24 scrambled symbols are then applied to the C2 encoder which generates four 8-bit parity symbols called Q words. The C2 encoder inserts the Q words between the 24 incoming symbols, so that at the output of the C2 encoder 28 symbols result
- between the C2 and the C1 decoder there are twenty-eight 8-bit delay lines with unequal delays. Due to the different delays, the sequence of the symbols is changed completely, according to a determined pattern
- the C1 encoder generates further four 8-bit parity symbols known as P words, resulting in a total of thirty-two 8-bit symbols.
- after the C1 encoder, the even words are subjected to a one-symbol delay, and all P and Q control words are inverted. The resultant sequenced thirty-two 8-bit symbols is called a **frame** and is a CIRC-encoded signal and is applied to the EFM modulator. On playback, the CIRC decoding circuit restores the original 16-bit samples which are then applied to the D/A converter.

The C2 encoder outputs twenty-eight 8-bit symbols for 24 symbols at its input: it is therefore called a (24, 28) encoder. The C1 encoder outputs 32 symbols for 28 symbols input: it is a (28, 32) encoder.

The bit rate at the output of the CIRC encoder is:

$$1.4112 \times \frac{32}{24} = 1.8816 \times 10^6 \text{ bit s}^{-1}$$

The Control Word

One 8-bit control word is added to every 32-symbol block of data from the encoder. The compact disc standard defines eight additional channels of information or **subcodes** that can be added to the music information; these subcodes are called P, Q, R, S, T, U, V and W. At the time of writing, only the P and Q subcodes are commonly used:

- the P subcode is a simple music track separator flag that is normally 0 (during music and in the lead-in track) but is 1 at the start of each selection. It can be used for simple search systems. In the lead-out track, it switches between 0 and 1 in a 2 Hz rhythm to indicate the end of the disc.
- the Q subcode is used for more sophisticated control purposes; it contains data such as track number and time.

  The other subcodes carry information relating to possible enhancements, such as text and graphics but will not be discussed here.

Each subcode word is 98 bits long: and, as each bit of the control word corresponds to each subcode (i.e., P, Q, R, S, T, U, V, W), a total of 98

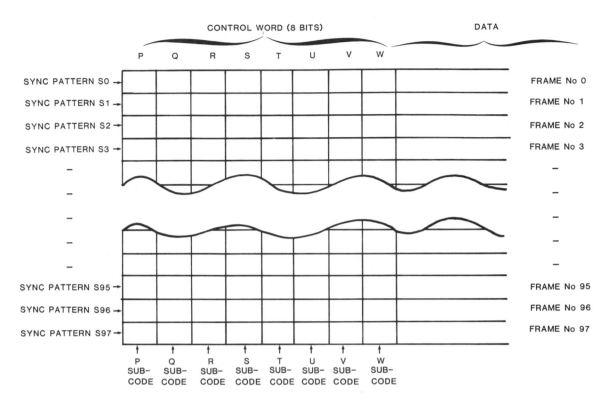

Figure 9.3 *Showing how one of each of the six subcode bits are present in every frame of information. A total of 98 frames must therefore be read to read all six subcode words*

complete data blocks or frames must be read from the disc to read each subcode word. This is illustrated in Figure 9.3.

After addition of the control word, the new data rate becomes:

$$1.811600 \times \frac{33}{32} = 1.9404 \times 10^6 \text{ bit s}^{-1}$$

The Q Subcode and its Usage

Figure 9.4a illustrates the structure of the 98-bit Q subcode word. The R, S, T, U, V, and W subcode words are similar. The first two bits are synchronizing bits, S0 and S1. They are necessary to allow the decoder to distinguish the control word in a block from the audio information, and always contain the same data.

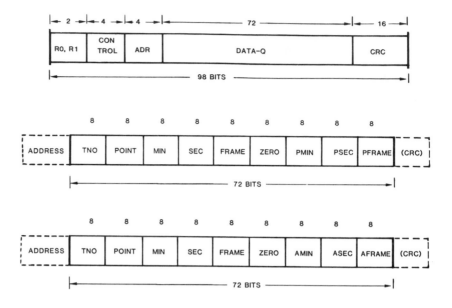

Figure 9.4 *Formats of data in the Q subcode (a) overall format (b) mode 1 data format in the lead-in track (c) mode 1 data format in music and lead-out tracks*

The next four bits are control bits, indicating the number of channels and pre-emphasis used, as follows:

0000   2 audio channels/no pre-emphasis
1000   4 audio channels/no pre-emphasis
0001   2 audio channels/with pre-emphasis
1001   4 audio channels/with pre-emphasis

Four address bits indicate the mode of the subsequent data to follow. For the Q subcode, three modes are defined.

At the end of the subcode word a 16-bit CRCC error-correction code, calculated on control, address and data information, is inserted. The CRCC uses the polynomial $P(X) = X^{16} + X^{12} + X^5 + 1$.

The three modes of data in Q subcode words are used to carry various information.

Mode 1 (address = 0001)

This is the most important mode, and the only one which is of use during normal playback. At least 9 out of 20 consecutive subcode words must carry data in mode-1 format. Two different situations are possible, depending whether the subcode is in the lead-in track or not.

When in the lead-in track, the data are in the format illustrated in Figure 9.4b. The 72-bit section comprises nine 8-bit parts:

- **TNO** – containing information relating to track number: two digits in BCD form (i.e., 2×4 bits). Is 00 during lead-in.
- **POINT/PMIN/PSEC/PFRAME** – containing information relating to the table of contents (TOC). They are repeated three times.

    POINT indicates the successive track numbers, while PMIN, PSEC and PFRAME indicate the starting time of that track. Furthermore, if POINT = A0, PMIN gives the physical track number of the first piece of music (PSEC and PFRAME are zero); if POINT = A1, PMIN indicates the last track on the disc, and if POINT = A2, the starting-point of the lead-out track is given in PMIN, PSEC and PFRAME.

    Table 9.2 shows the encoding of the TOC on a disc which contains 6 pieces of music.

- **ZERO** – eight bits, all 0.

    In music and lead-out tracks, data are in the format illustrated in Figure 9.4c. The 72-bit section now comprises:

- **TNO** – current track number: two digits in BCD form (01 to 99).
- **POINT** – index number within a track: two digits in BCD form (01 to 99). If POINT = 00 it indicates a pause in between tracks.
- **MIN/SEC/FRAME** – indicates running time within a track: each part, of digits in BCD form. There are 75 frames in a second (00 to 74). Time is counted down during a pause, with a value zero at the end of the pause. During lead-in and lead-out tracks, the time increases.
- **AMIN/ASEC/AFRAME** – indicates the running time of the disc in same format as above. At the start of the programme area, it is set to zero.
- **ZERO** – eight bits, all 0.

Figure 9.5 shows a timing diagram of P subcode and Q subcode status during complete reading of a disc containing four selections (of which selections three and four fade out and in consecutively without an actual pause).

Mode 2 (address = 0010)

If mode 2 data are present, at least 1 out of 100 successive subcode words must contain it. It is of importance only to the manufacturer of the disc, containing the disc catalogue number. The 98-bit, Q subcode word in mode 2 is shown in Figure 9.6. Structure is similar to that of mode 1, with the following differences:

- **N1 to N13** – catalogue number of the disc expressed in 13 digits of BCD, according to the UPC/EAN standard for bar coding. The catalogue

Table 9.2 *Table of contents (TOC) information, on a compact disc with six pieces of music*

Frame number	POINT	PMIN, PSEC, PFRAME
n	01	00, 02, 32
n + 1	01	00, 02, 32
n + 2	01	00, 02, 32
n + 3	02	10, 15, 12
n + 4	02	10, 15, 12
n + 5	02	10, 15, 12
n + 6	03	16, 28, 63
n + 7	03	16, 28, 63
n + 8	03	16, 28, 63
n + 9	04	16, 28, 63
n + 10	04	16, 28, 63
n + 11	04	16, 28, 63
n + 12	05	16, 28, 63
n + 13	05	16, 28, 63
n + 14	05	16, 28, 63
n + 15	06	49, 10, 33
n + 16	06	49, 10, 33
n + 17	06	49, 10, 33
n + 18	A0	01, 00, 00
n + 19	A0	01, 00, 00
n + 20	A0	01, 00, 00
n + 21	A1	06, 00, 00
n + 22	A1	06, 00, 00
n + 23	A1	06, 00, 00
n + 24	A2	52, 48, 41
n + 25	A2	52, 48, 41
n + 26	A2	52, 48, 41
n + 27	01	00, 02, 32
n + 28	01	00, 02, 32
.	.	.
.	.	.
.	.	.

number is constant for any one disc. If no catalogue number is present, N1 to N13 are all zero, or mode 2 subcode words may not even appear.

● **ZERO** – these 12 bits are zero.

Mode 3 (address = 0111)

Like mode 2 data, if mode 3 is present, at least 1 out of 100 successive subcode words will contain it.

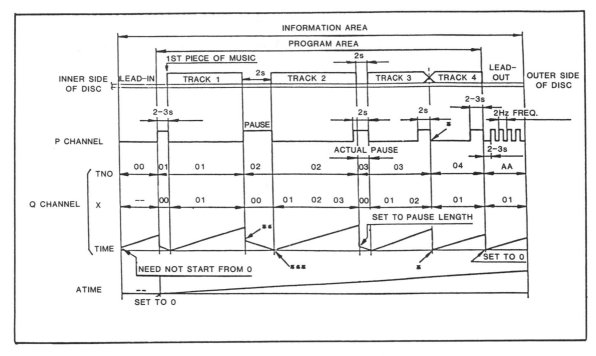

Figure 9.5 *Timing diagram of P and Q subcodes*

Figure 9.6 *Q subcode format with mode 2 data*

Mode 3 is used to assign each selection with a unique number, according to the 12-character International Standard Recording Code (ISRC), defined in DIN-31-621.

If no ISRC number is assigned, mode 3 subcode words are not present. During lead-in and lead-out tracks, mode 3 subcode words are not used, and the ISRC number must only change immediately after the track number (TNO) has been changed.

The 98-bit, Q subcode word in mode 3 is shown in Figure 9.7. Structure is similar to that of mode 1, with the following differences:

● **I1 to I12** – the 12 characters of the selection's ISRC number. Characters I1 and I2 give the code corresponding to country. Characters I3 to I5 give a

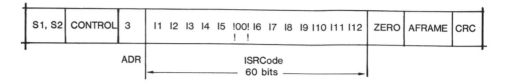

Figure 9.7 *Q subcode format with mode 3 data*

Table 9.3 *Format of characters I1 to I5 in the ISRC code*

Character	Binary	Octal	Character	Binary	Octal
0	000000	00	I	011001	31
1	000001	01	J	011010	32
2	000010	02	K	011011	33
3	000011	03	L	011100	34
4	000100	04	M	011101	35
5	000101	05	N	011110	36
6	000110	06	O	011111	37
7	000111	07	P	100000	40
8	001000	10	Q	100001	41
9	001001	11	R	100010	42
A	010001	21	S	100011	43
B	010010	22	T	100100	44
C	010011	23	U	100101	45
D	010100	24	V	100110	46
E	010101	25	W	100111	47
F	010110	26	X	101000	50
G	010111	27	Y	101001	51
H	011000	30	Z	101010	52

code for the owner. Characters I6 and I7 give the year of recording. Characters I8 to I12 give the recording's serial number.

Characters I1 to I5 are coded in a 6-bit format according to Table 9.3, while characters I6 to I12 are 4-bit BCD numbers.

- **00** – these two bits are zero.
- **ZERO** – these four bits are zero.

EFM Encoding

EFM, or **eight-to-fourteen modulation**, is a technique which converts each 8-bit symbol into a 14-bit symbol, with the purpose of aiding the recording and playback procedure by reducing required bandwidth, reducing the

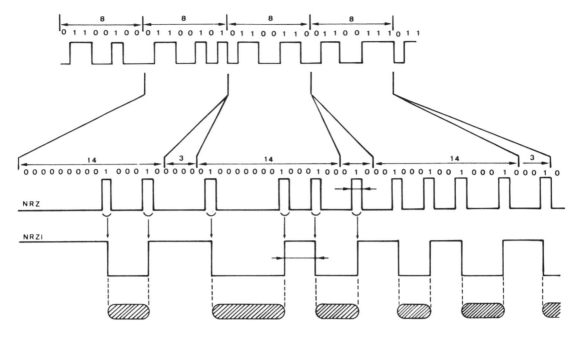

Figure 9.8 *Timing diagram of EFM encoding and merging bits*

signal's DC content, and adding extra synchronization information. A timing diagram of signals in this stage of CD encoding is given in Figure 9.8.

The procedure is to use 14-bit codewords to represent all possible combinations of the 8-bit code. An 8-bit code represents 256 (i.e., 2^8) possible combinations, as shown in Table 9.4. A 14-bit code, on the other hand, represents 16,384 (i.e., 2^{14}) different combinations, as shown in Table 9.5. Out of the 16,384 14-bit codewords, only 256 are selected, having combinations which aid processing of the signal.

Table 9.4 *An 8-bit code*

MSB	7SB	6SB	5SB	4SB	3SB	2SB	LSB	N⁰
0	0	0	0	0	0	0	0	0
0	0	0	0	0	0	0	1	1
0	0	0	0	0	0	1	0	2
—	—	—	—	—	—	—	—	—
—	—	—	—	—	—	—	—	—
1	1	1	1	1	1	1	0	254
1	1	1	1	1	1	1	1	255

Table 9.5 *A 14-bit code*

MSB	13SB	12SB	11SB	10SB	9SB	8SB	7SB	6SB	5SB	4SB	3SB	2SB	LSB	N°
0	0	0	0	0	0	0	0	0	0	0	0	0	0	0
0	0	0	0	0	0	0	0	0	0	0	0	0	1	1
0	0	0	0	0	0	0	0	0	0	0	0	1	0	2
–	–	–	–	–	–	–	–	–	–	–	–	–	–	–
–	–	–	–	–	–	–	–	–	–	–	–	–	–	–
–	–	–	–	–	–	–	–	–	–	–	–	–	–	–
1	1	1	1	1	1	1	1	1	1	1	1	1	0	16382
1	1	1	1	1	1	1	1	1	1	1	1	1	1	16383

Table 9.6 *Examples of 8-bit to 14-bit encoding*

8-bit word	14-bit word
0 0 0 0 0 0 1 1	0 0 1 0 0 1 0 0 0 0 0 0 0 0
0 1 0 0 1 1 1 0	0 1 0 0 0 0 0 1 0 0 1 0 0 0
1 0 1 0 1 0 1 0	1 0 0 1 0 0 0 1 0 0 0 1 0 0
1 1 1 1 0 0 1 0	0 0 0 0 0 0 1 0 0 0 1 0 0 1

For instance, by choosing codewords which give low numbers of individual bit inversions (i.e., 1 to 0, or 0 to 1) between consecutive bits, the bandwidth is reduced. Similarly, by choosing codewords with only limited numbers of consecutive bits with the same logic level, overall DC content is reduced.

A ROM-based look-up table, say, can then be used to assign all 256 combinations of the 8-bit code to the 256 chosen combinations within the 14-bit code. Some examples are listed in Table 9.6.

In addition to EFM modulation, three extra bits, known as **merging bits**, are added to each 14-bit symbol, with the purpose of further lowering DC content of the signal. Exact values of the merging bits depend on the adjacent symbols.

Finally, the data bits are changed from NRZ into NRZI (non-return to zero inverted) format, by converting each positive-going pulse of the NRZ signal into a single transition. The resultant signal has a minimum length of 3T (i.e., three clock periods), and a maximum of 11T (i.e., 11 clock periods), as shown in Figure 9.9.

Bit rate is now:

$$1.9404 \times \frac{17}{8} = 4.12335 \times 10^6 \text{ bit s}^{-1}$$

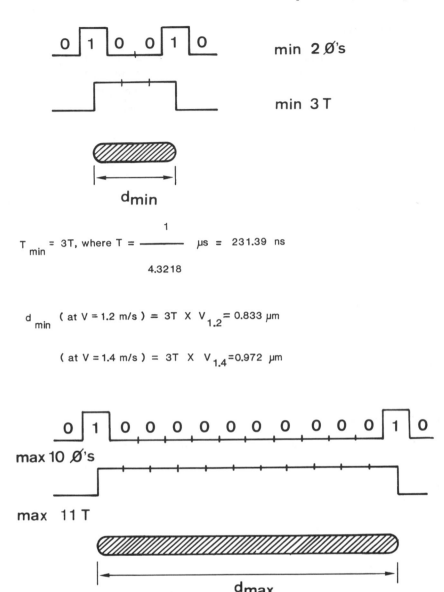

$$T_{min} = 3T, \text{ where } T = \frac{1}{4.3218} \ \mu s = 231.39 \text{ ns}$$

$$d_{min} \ (\text{at } V = 1.2 \text{ m/s}) = 3T \times V_{1.2} = 0.833 \ \mu m$$

$$(\text{at } V = 1.4 \text{ m/s}) = 3T \times V_{1.4} = 0.972 \ \mu m$$

$$d \ max \ (\text{at } V = 1.2 \text{ m/s}) = 11 T \times V_{1.2} = 3.05 \ \mu m$$

$$(\text{at } V = 1.4 \text{ m/s}) = 11 T \times V_{1.4} = 3.56 \ \mu m$$

Figure 9.9 *Minimum and maximum pit length*

The Sync Word

To the signal, comprising 33 symbols of 17 bits (i.e., a total 561 bits) a sync word and its three merging bits are added, giving 588 bits in total (Figure 9.10). Sync words have two main functions: (1) they indicate the start of each frame (2) sync word frequency is used to control the player's motor speed.

The 588 bit long signal block is known as an information **frame**.

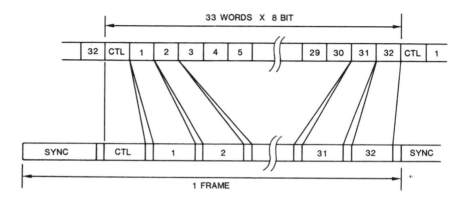

Figure 9.10 *Adding the sync word*

Final Bit Rate

The final bit rate, recorded on the CD, consequently becomes:

$$4.12335 \times \frac{588}{561} = 4.3218 \times 10^6 \text{ bit s}^{-1}$$

The frame frequency F_{frame}, is:

$$\frac{4.3218}{588} = 7350 \text{ Hz}$$

And, as subcodes are in blocks of 98 frames, the subcode frequency F_{sc}, is:

$$= \frac{7350}{98} = 75 \text{ Hz}$$

Playing time is calculated by counting blocks of subcode (i.e., 75 blocks = 1 second). A 60-minute long CD contains consequently:

$$60 \times 60 \times 7350 = 26{,}460{,}000 \text{ frames}$$

As each frame comprises $33 \times 8 = 264$ bits of information, a one-hour long CD actually contains 6,985,440,000 bits of information, or 873,180,000 bytes! Of this, the subcode area contains some 25.8 kbyte (200 Mbits). This gigantic data storage capacity of the CD medium is also used for more general purposes on the CD-ROM (compact disc read only memory), which is derived directly from the audio CD.

10
Opto-electronics and the Optical Block

As the compact disc player uses a laser beam to read the disc, we will sketch some basic principles of opto-electronics – the technological marriage of the fields of optics and electronics. The principles are remarkably diverse, involving such topics as the nature of optical radiation, the interaction of light with matter, radiometry, photometry and the characteristics of various sources and sensors.

The Optical Spectrum

By convention, electromagnetic radiation is specified according to its wavelength (λ). The frequency of a specific electromagnetic wavelength is given by:

$$f = \frac{c}{\lambda}$$

where f is frequency in Hz, c is velocity of light ($3 \times 10^8 \, \text{ms}^{-1}$), λ is wavelength in m.

The optical portion of the electromagnetic spectrum extends from 10 nm to 10^6 nm and is divided into three major categories: ultraviolet (UV), visible and infrared (IR).

Ultraviolet (UV) are those wavelengths, falling below the visible spectrum and above x-rays. UV is classified according to its wavelength as extreme or shortwave UV (10 to 200 nm), far (200 to 300 nm) and near or long-wave UV (300 to 370 nm).

Visible are those wavelengths between 370 to 750 nm and they can be perceived by the human eye. Visible light is classified according to the various

Photo 10.1 *Optical block of a compact disc player*

colours its wavelengths elicit in the mind of a standard observer. The major colour categories are: violet (370 to 455 nm), blue (456 to 492 nm), green (493 to 577 nm), yellow (578 to 597 nm), orange (598 to 622 nm) and red (623 to 750 nm).

Infrared (IR) are those wavelengths above the visible spectrum and below microwaves. IR is classified according to its wavelength as near (750 to 1500 nm), middle (1600 to 6000 nm), far (6100 to 40000 nm) and far-far (41000 to 10^6 nm).

Interaction of Optical Waves with Matter

An optical wave may interact with matter by being reflected, refracted, absorbed or transmitted. The interaction normally involves two or more of these effects.

Reflection

Some of the optical radiation impinging upon any surface is reflected away from the surface. Amount of reflection varies according to the properties of the surface and the wavelength and in real circumstances may range from more than 98% to less than 1% (a **lampblack** body). Reflection from a surface may be diffuse, specular or a mixture of both.

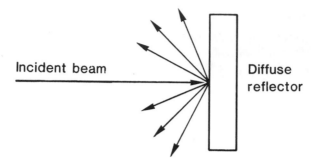

Figure 10.1 *Reflection of light from a diffuse reflector*

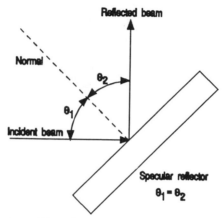

Figure 10.2 *Reflection of light from a specular reflector*

A diffuse reflector has a surface which is rough when compared to the wavelength of the impinging radiation (Figure 10.1).

A specular, sometimes called regular, reflector, on the other hand has a surface which is smooth when compared to the wavelength of the impinging radiation. A perfect specular reflector will thus reflect an incident beam without altering the divergency of the beam.

A narrow beam of optical radiation impinging upon a specular reflector obeys two rules, illustrated in Figure 10.2.

1 the angle of reflection is equal to the angle of incidence
2 the incident ray and the reflected ray lie in the same plane as a normal line extending perpendicularly from the surface.

Absorption

Some of the optical radiation impinging upon any substance is absorbed by the substance. Amount of absorption varies according to the properties of the substance and the wavelength and in real circumstances any range from less than 1% to more than 98%.

Transmission

Some of the optical radiation impinging upon a substance is transmitted into the substance. The penetration depth may be shallow (transmission = 0) or deep (transmission more than 75%).

The reflection (ρ), the absorption (α) and the transmission (σ) are related in the expression:

$$\rho + \alpha + \sigma = 1$$

Refraction

A ray of optical radiation passing from one medium to another is bent at the interface of the two mediums if the angle of incidence is not 90°.

The index of refraction n, is the sine of the angle of incidence divided by the sine of the angle of refraction, as illustrated in Figure 10.3. Refractive index varies with wavelength and ranges from 1.0003 to 2.7.

Optical Components

Optical components are used both to manipulate and control optical radiation and to provide optical access to various sources and sensors.

Glass is the most common optical material at visible and near-infrared wavelengths, but other wavelengths require more exotic materials such as calcium aluminate glass (for middle infrared), lithium fluoride (for UV).

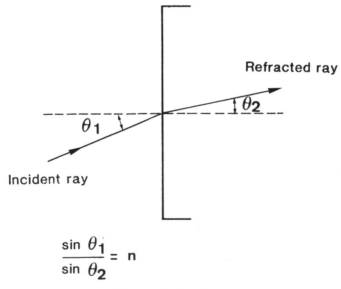

$$\frac{\sin \theta_1}{\sin \theta_2} = n$$

Figure 10.3 *Illustrating the index of refraction*

The thin simple lens

Figure 10.4 shows how an optical ray passes through a thin simple lens.

A lens may be either positive (converging) or negative (diverging). The focal point of a lens is that point at which the image of an infinitely distant point source is reproduced.

Lens (diameter D)

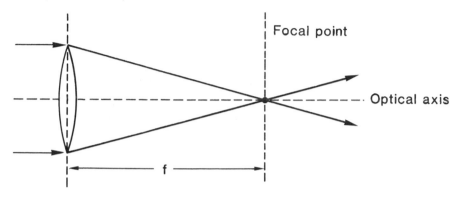

Figure 10.4 *Showing how light is refracted through a thin simple lens*

Both the source and the focal point lie on the lens axis. The focal length (f) is the distance between the lens and the focal point. The f/number of a lens defines its light-collecting ability and is given by:

$$f/number = \frac{f}{D}$$

where f is the focal length, D is the diameter of the lens.

A small f/number denotes a large lens diameter for a specified focal length and a higher light-collecting ability than a large f/number.

Numerical aperture (NA) is a measure of the acceptance angle of a lens and is given by:

$$NA = n \sin \theta$$

where n is the refractive index of the object or image medium (for air, n = 1), θ is half the maximum acceptance angle (shown in Figure 10.5). The relation of the focal length (f) to the distances between the lens and the object being imaged (s) and the lens and the focused image (s') is given by the gaussian form of the thin lens equation:

$$1/s + 1/s' = 1/f$$

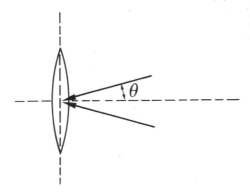

Figure 10.5 *Acceptance angle of a lens*

The combined focal length for two thin lenses in contact or close proximity and having the same optical axis is given by:

$$1/f = 1f_1 + 1/f_2$$

The relationship between the focal length (f) and the refractive index (n) is:

$$1/f = (n - 1)(1/r_1 - 1/r_2)$$

where r_1 is the radius on the left lens surface, r_2 is the radius on the right lens surface.

The cylindrical lens

A cylindrical lens is a section of a cylinder and therefore magnifies in only one plane. An optical beam which enters the cylindrical lens of Figure 10.6 is focused only in the horizontal plane. As a result the cross-section of the beam after passing through the lens is elliptic, with the degree varying according to the distance from the lens. By detecting the elliptic degree, a useful measure of whether or not a beam is focused on a surface can be made.

The prism

A prism is an optically transparent body used to refract, disperse or reflect an optical beam. The simplest prism is the right-angle prism, shown in Figure 10.7.

An optical ray perpendicularly striking one of the shorter faces of the prism is totally internally reflected at the hypotenuse, undergoes a 90° deviation, and emerges from the second shorter face.

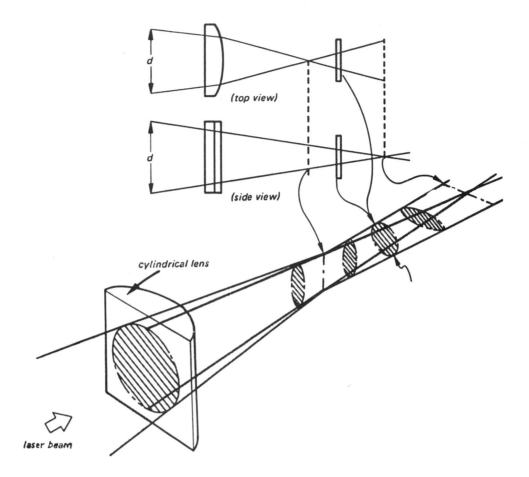

(top view)

(side view)

cylindrical lens

laser beam

Figure 10.6 *Operation of a cylindrical lens*

Operation of a totally internally reflecting prism is dependent upon the fact that a ray impinging upon the surface of a material having a refractive index (n) smaller than the refractive index (n') of the medium in which the ray is propagating, will be totally internally reflected when the angle of incidence is greater than a certain critical angle (θ_c), given by:

$$\sin \theta_c = \frac{n'}{n}$$

The collimator

The combination of two simple lenses is commonly used to increase the diameter of a beam while reducing its divergence. Such a **collimator** is shown in Figure 10.8.

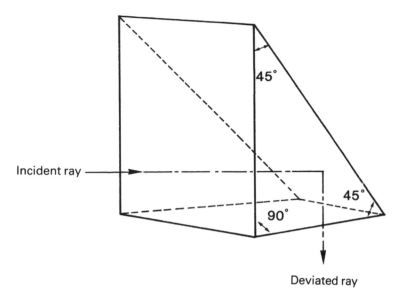

Figure 10.7 *Right-angle prism*

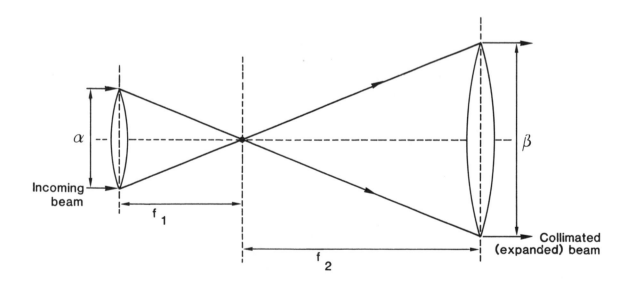

$$\frac{\alpha}{\beta} = \frac{f_1}{f_2}$$

Figure 10.8 *Principle of a collimator*

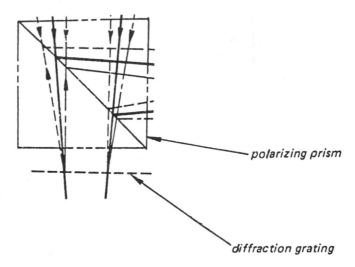

Figure 10.9 *A diffraction grating used with a prism (part of a CD player optical pick-up)*

Diffraction gratings

Diffraction gratings are used to split optical beams and usually comprise a thin plane parallel plate, one surface of which is coated with a partially reflecting film of thin metal with thin slits. Figure 10.9 shows a diffraction grating used with a prism.

The slits are spaced only a few wavelengths apart. When the beam passes through the grating it diffracts at different angles, and appears as a bright main beam with successively less intensive side beams, as shown in Figure 10.10.

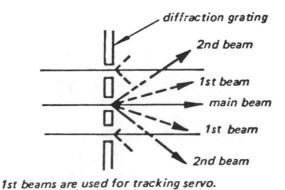

1st beams are used for tracking servo.

Figure 10.10 *Light passing through a diffraction grating*

1/4 wave plate

When linear polarized light is passed through an anisotropic crystal (Figure 10.11), the polarization plane will be contorted during the traverse of the crystal. The thickness d, of the crystal required to obtain a contortion of more than 180°, is equal to one wavelength of the light. In order to obtain a contortion of more than 45°, only 1/4 of d is therefore required, and a crystal with this thickness is used in an optical pick-up in a CD player, and is called a **1/4 wave-plate**, **quarter wave plate**, or **QWP**.

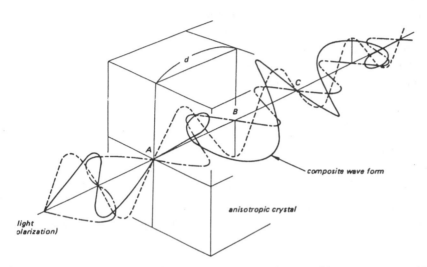

Figure 10.11 *Light passing through an anisotropic crystal becomes rotated in polarization*

The Injection Laser Diode (ILD)

The basic operation of any laser consists of pumping the atoms into a stimulated state, from which electrons can escape and fall to the lower energy state by giving up a photon of the appropriate energy. In a solid-state laser, the input energy populates some of the usually unpopulated bands. When a photon of energy equal to the band gap of the crystal passes through, it stimulates other excited photons to fall in step with it. Thus a small **priming** signal will emerge with other coherent photons.

The laser can be operated in a continuous wave (CW) oscillator mode if the ends of the laser path are optically flat, parallel and reflective, so forming an optical resonant cavity. A spontaneously produced photon will rattle back and forth between the ends, acquiring companions due to stimulated emission.

This stimulated emission can be like an avalanche, completely draining the high-energy states in a rapid burst of energy. On the other hand, certain types of lasers can be produced which will operate in an equilibrium condition, giving up photons at just the input energy pumping rate. The **injection laser diode** (ILD) is such a device.

The material in an injection laser diode is heavily doped so that under forward bias, the region near the junction has a very high concentration of holes and electrons. This produces the inversion condition with a large population of electrons in a high-energy band and a large population of holes in the low-energy band. In this state, the stimulated emission of photons can overcome photon absorption and a net light flux will result.

In operation, the forward current must reach some threshold: beyond which the laser operates with a single 'thread' of light, and the output is relatively stable but low. As the current is increased, light output increases rapidly.

ILD characteristic is highly temperature-sensitive. A small current variation or a modest temperature change can cause the output to rise so rapidly that it destroys the device. A photodiode, monitoring the light output, is commonly used in a feedback loop to overcome this problem.

Like the LED, the ILD must be driven from a current source, rather than a voltage source, to prevent thermal runaway. With a voltage source, as the device junction begins to warm the forward voltage drop decreases, which

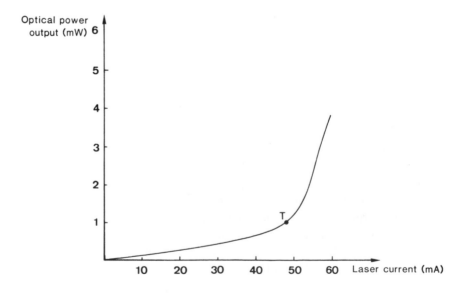

Figure 10.12 *Characteristic of a typical ILD*

tends to increase the current, in turn decreasing the forward voltage drop, and so on until the current tends towards infinity and the device is destroyed.

In addition, the ILD has a tendency to deteriorate with operation. Deterioration is greatly accelerated when operating the ILD outside of its optimum limits. A typical ILD characteristic is shown in Figure 10.12.

The ILD with a double heterostructure (DH)

A very narrow P-N junction layer of GaAs semiconductor is sandwiched between layers of AlGaAs.

The properties of the outer semiconductor layers confine the electrons and holes to the junction layer, leading to an inverted population at a low input current.

Quantum aspects of the laser

A laser (light amplification by stimulated emission of radiation) is a maser (microwave amplification of stimulated emission of radiation), operating at optical frequencies.

Because the operations of masers and lasers are dependent upon quantum processes and interactions, they are known as quantum electronic devices. The laser, for example, is a quantum amplifier for optical frequencies, whereas the maser is a quantum amplifier for microwave frequencies.

Masers and lasers utilize a solid or gaseous active medium, and their operations are dependent on Planck's law:

$$\Delta E = E_2 - E_1 = hf$$

An atom can make discontinuous jumps or transitions from one allowed energy level (E_2) to another (E_1), accompanied by either the emission or absorption of a photon of electromagnetic radiation at frequency f.

The energy difference E is commonly expressed in electronvolts (abbreviation: eV) and:

$$1\,eV = 1.6 \times 10^{-19}\,J$$

The constant h is Planck's constant and equals $6.6262 \times 10^{-34}\,Js$. The frequency f (in hertz) of the radiation and the associated wavelength λ (in metres) are related by the expression:

$$\lambda = \frac{c}{f}$$

where c (the velocity of light) is $3 \times 10^8\,ms^{-1}$.

Example:

$$\Delta E = E_2 - E_1 = 1.6\,\text{eV}$$
$$= 1.6 \times 1.6 \times 10^{-19}\,\text{J}$$
$$= 2.5632 \times 10^{-19}\,\text{J}$$

But, ΔE also equals:

$$hf = h\frac{c}{\lambda}$$

So:

$$\lambda = \frac{hc}{\Delta E}$$

$$= \frac{6.6262 \times 10^{-34} \times 3 \times 10^8}{2.5632 \times 10^{-19}}\,\text{m}$$

$$= 775 \times 10^{-9}\,\text{m}$$

$$= 775\,\text{nm}$$

TOP: T-type Optical Pick-up

The schematic of an optical pick-up of the three-beam type used in CD players is shown in Figure 10.13. This drawing shows the different optical elements composing a pick-up and indicates the laser beam route through the unit.

The laser beam is generated by the laser diode. It passes through the diffraction grating, generating two secondary beams called **side beams**, which are used by a tracking servo circuit to maintain correct tracking of the disc.

The beam enters a polarizing prism (called a **beam splitter**) and only the vertical polarized light passes. The light beam, still divergent at this stage, is converged into a parallel beam by the collimation lens and passed through the 1/4 wave plate where the beam's polarization plane is contorted by 45°. The laser beam is then focused by a simple lens onto the pit surface of the disc. The simple lens is part of a servo-controlled mechanism known as a **2-axis device**.

The beam is reflected by the disc mirrored surface, converged by the 2-axis device lens into a parallel beam, and re-applied to the 1/4 wave plate. Again, the polarization plane of the light is contorted by 45°, so the total amount of contortion becomes 90°, i.e., the vertically polarized laser beam has been twisted to become horizontally polarized.

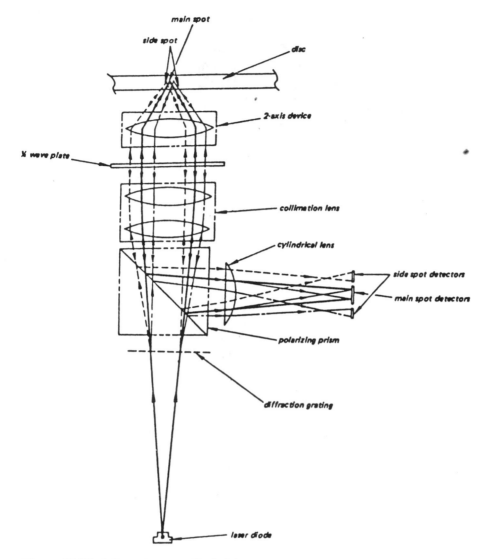

Figure 10.13 *A three-beam optical pick-up*

After passing through the collimation lens the laser beam is converged onto the slating surface of the polarizing prism. Now polarized horizontally, the beam is reflected internally by the slanted surface of the prism towards the detector part of the pick-up device.

In the detector part, the cylindrical lens focuses the laser beam in only one plane, onto six photo-detectors in a format shown in Figure 10.14, i.e., four main spot detectors (A, B, C and D) and two side spot detectors (E and F), enabling read-out of the pit information from the disc.

The T-type optical pick-up (TOP) is shown in detail in Figure 10.15.

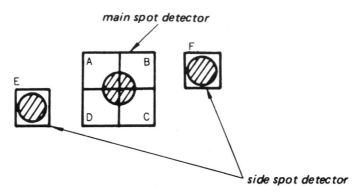

Figure 10.14 *Showing relative positions of the six photo-detectors of a CD optical pick-up*

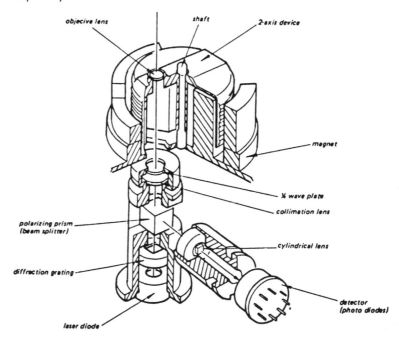

Figure 10.15 *T-type optical pick-up (TOP)*

FOP: Flat-type Optical Pick-up

Later models use a flat-type optical pick-up (FOP) shown in Figure 10.16. In the FOP a non-polarizing prism is used and no 1/4 wave plate. The non-polarizing prism is a half mirror which reflects half of the incident light and lets pass the other half (Figure 10.17).

This means that half of the light passes through the prism and returns after being reflected by the mirror. This secondary beam, 50% of the original, is

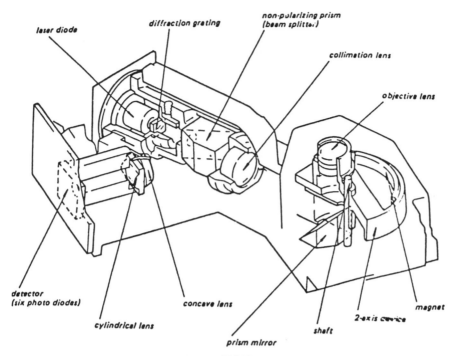

Figure 10.16 *Flat-type optical pick-up (FOP)*

reflected again for 50% and passed through for 50%, so that the resulting light beam intensity is $1/2 \times 1/2 = 1/4$ of the original beam (Figure 10.18).

The new principle enables the elimination of the influence of double refraction (i.e. the change of the deflection angle when reflected by the disc surface), caused by discs with mirror impurities.

2-axis device

Optical pick-ups contain an actuator for objective lens position control. The compact disc player, due to the absence of any physical contact between the disc and the pick-up device, has to contain auto-focus and auto-tracking functions.

These functions are performed by the focus and tracking servo circuits via the 2-axis device, enabling a movement of the objective lens in two axes: vertically for focus correction and horizontally for track following. Figure 10.19 shows such a 2-axis device construction.

The principle of operation is that of the moving coil in a magnetic field. Two coils, the **focus coil** and the **tracking coil**, are suspended between magnets (Figure 10.20) creating two magnetic fields. A current through either coil, due to the magnetic field, will cause the coil to be subjected to a force, moving the coil in the corresponding direction.

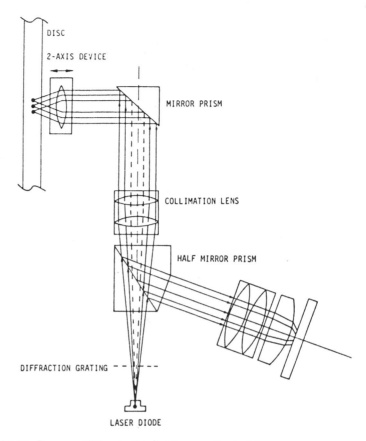

Figure 10.17 *Beam splitting in the flat-type optical pick-up*

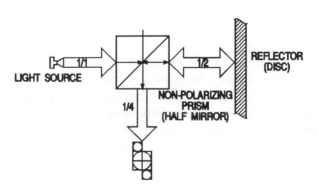

Figure 10.18 *Light distribution in FOP*

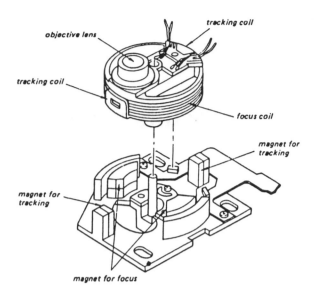

Figure 10.19 *Construction of a 2-axis device, used to focus the laser beam onto the surface of a compact disc*

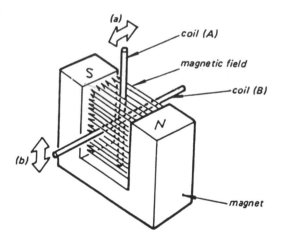

Figure 10.20 *Operating principle of a 2-axis device*

11
The Servo Circuits in CD Players

A feedback control circuit is one in which the value of a controlled variable signal is compared with a reference value. The difference between these two signals generates an actuating error signal which may then be applied to the control elements in the control system. Principle is shown in Figure 11.1. The amplified actuating error signal is said to be fed back to the system, thus tending to reduce the difference. Supplementary power for signal amplification is available in such systems.

The two most common types of feedback control systems are **regulators** and **servo circuits**. Fundamentally, both systems are similar, but the choice of systems depends on the nature of reference inputs, the disturbance to which the control is subjected, and the number of integrating elements in the control.

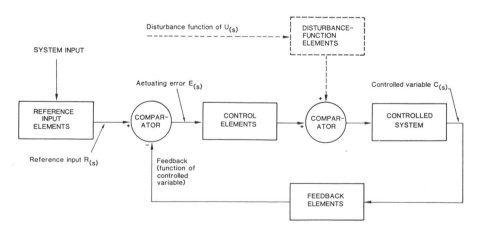

Figure 11.1 *Feedback control system: block diagram*

Regulators are designed primarily to maintain the controlled variable or system output very nearly equal to a desired value in the presence of output disturbances. Generally, a regulator does not contain any integrating elements. An example of a regulator is shown in Figure 11.2, a stabilized power supply with series regulator.

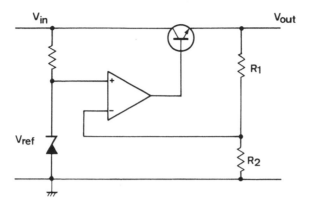

Figure 11.2 *A voltage regulator*

The non-inverting input of a comparator is connected to a reference voltage (V_{ref}), and a fraction of the output voltage V_{out} is fed back to the comparator's inverting input. Closed-loop gain of this circuit equals:

$$G = \frac{R1 + R2}{R2}$$

and, the output voltage equals:

$$V_{out} = G . V_{ref}$$

A servo circuit, on the other hand, is a feedback control system in which the controlled variable is mechanical, usually a displacement or a velocity. Ordinarily in a servo circuit, the reference input is the signal of primary importance; load disturbances, while they may be present, are of secondary importance. Generally, one or more integrating elements are contained in the forward transfer function of the servo circuit.

An example of a servo circuit is shown in Figure 11.3, where a motor is driven at a constant speed. This circuit is a phase-locked system consisting of a phase-frequency detector, an amplifier with a filter, a motor and an encoder. The latter is a device which emits a number of pulses per revolution of the motor shaft. Therefore, the frequency of the encoder signal is directly proportional to the motor speed.

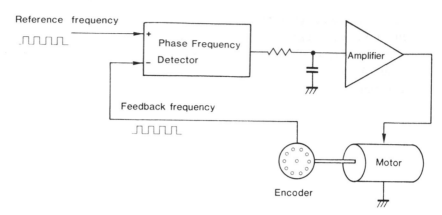

Figure 11.3 *Possible motor speed control servo circuit*

Objective of the system is to synchronize the feedback frequency with the reference frequency. This is done by comparing the two signals and correcting the motor velocity according to any difference in frequency or phase.

Summary of the Servo Circuits in a CD Player

For this explanation, the Sony CDP-101 CD player is used as an example. It uses four distinct servo circuits, as shown in Figure 11.4. These are:

1 **focus servo circuit**: this servo circuit controls vertical movement of the 2-axis device and guarantees that the focal point of the laser beam is precisely on the mirror surface of the compact disc
2 **tracking servo circuit**: this circuit controls the horizontal movement of the 2-axis device and forces the laser beam to follow the tracks on the compact disc
3 **sled servo circuit**: this circuit drives the sled motor which moves the optical block across the compact disc
4 **disc motor servo circuit**: this circuit controls the speed of the disc motor, guaranteeing that the optical pick-up follows the compact disc track at a constant linear velocity.

The optical pick-up is the source of the feedback signals for all four servo circuits.

The Focus Servo Circuit

Detection of the correct focal points

The reflected laser beam is directed to the main spot detector (Figure 11.5a), an array of four photodiodes, labelled A, B, C and D. When the focus is OK,

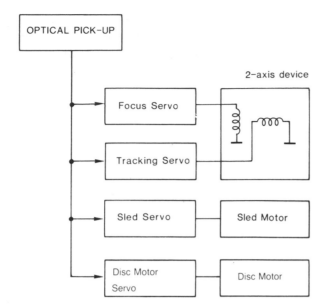

Figure 11.4 *Servo circuits in the CDP-101 compact disc player*

the beam falls equally on the four diodes, and the focus error signal: $(A + C) - (B + D)$ is zero.

On the other hand, when the beam is out of focus (Figures 11.5 b and c), an error signal is generated, because the beam passes through a cylindrical lens, which makes the beam elliptic in shape (Figure 11.6). The resultant focus error signal from the main spot detector: $(A + C) - (B + D)$ is therefore not zero.

The focus search circuit

When a disc is first loaded in the player the distance between the 2-axis device and the disc is too large: the focus error signal is zero (as shown in Figure 11.7a) and the focus servo circuit is inactive.

Therefore, a focus search circuit is used which, after the disc is loaded, moves the 2-axis device slowly closer to the disc. Outputs of the four photo-diodes are combined in a different way (i.e., $A + B + C + D$) to form a radio frequency (RF) signal, which represents the data bits read from the disc. When the RF signal exceeds a threshold level (Figure 11.7b), the focus servo is enabled and now controls the 2-axis device for a zero focus error signal.

The Tracking Servo Circuit

Figure 11.8 shows the three possible tracking situations as the optical pick-up follows the disc track. In Figure 11.8a and b the main spot detector is not

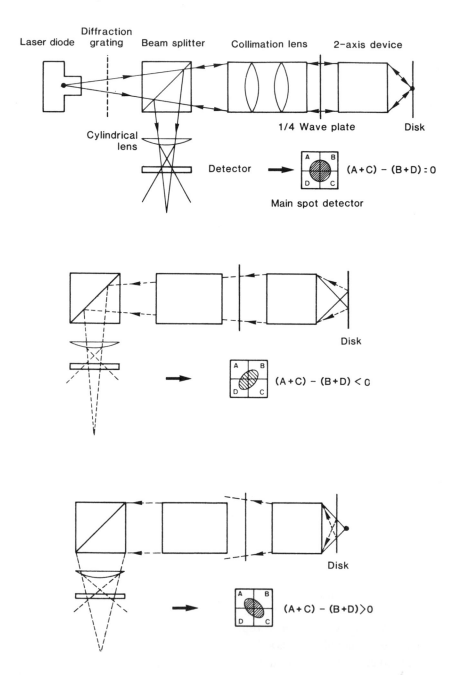

Figure 11.5 *Detection of correct focus (a) arrangement of optical pick-up: focus is correct (b) and (c) focus is not correct*

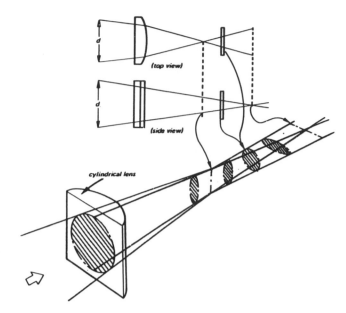

Figure 11.6 *Showing how elliptical beams are produced by the cylindrical lens, when the optical pick-up is out of focus*

correctly tracked, and so one or other of the side spot detectors gives a large output signal as the pit is traversed. In Figure 11.8c, on the other hand, the main spot detector is correctly tracked, and both side spot detectors give small output signals.

Side spot detectors consist of two photo-diodes (E and F) and generate a tracking error signal:

$$TE = E - F$$

The tracking servo acts in such a way that the tracking error signal is as small as possible, i.e., the main spot detector is exactly on the pits of the track. Tracking error, focus error, and resultant focus signals are shown, derived from the optical pick-up's photodiode detectors, in Figure 11.9.

The Sled Servo Motor

The 2-axis device allows horizontal movement over a limited number of tracks, giving a measure of fine tracking control. Another servo circuit, called the sled servo circuit, is used to move the complete optical unit across the disc for coarse tracking control. It uses the same tracking error signal as the tracking servo of the 2-axis device.

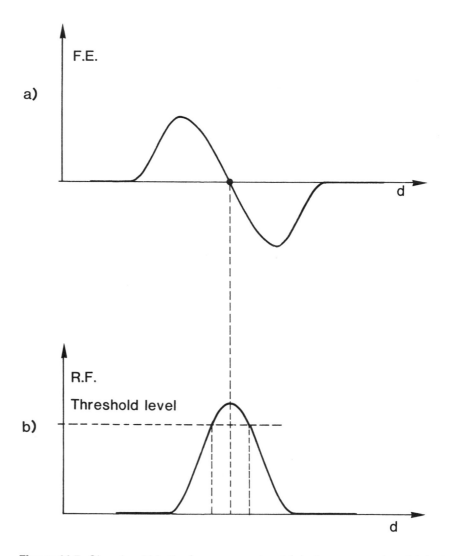

Figure 11.7 *Signals within the focus servo circuit (a) a focus error signal detects when the pick-up is in focus by means of combining the photo-detector outputs as $(A+C)-(B+D)$: zero voltage means focus has·been obtained (b) a radio frequency signal, obtained by combining the photo-detector outputs as $(A+B+C+D)$, must exceed a threshold level before the focus servo is activated*

However, the output of the tracking servo circuit is linearly related to the tracking error signal, whereas the output of the sled servo circuit has a built-in hysteresis: Only when the TE signal exceeds a fixed threshold level does the sled servo drive the sled motor. Tracking error and sled motor drive signals are shown in Figure 11.10.

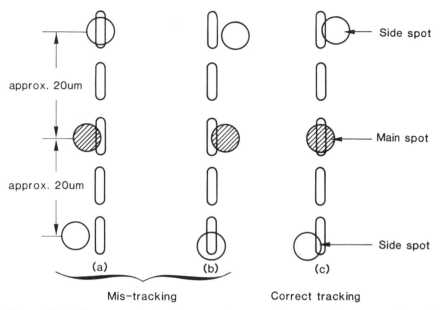

Figure 11.8 *Three possible tracking situations (a) and (b) mis-tracking (c) correctly tracking*

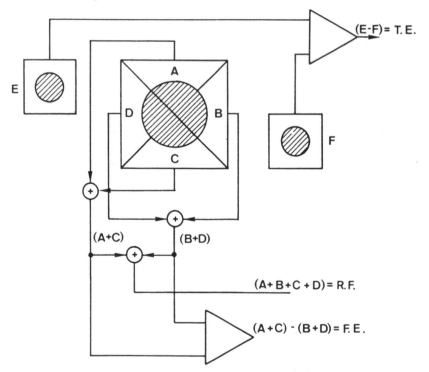

Figure 11.9 *Showing how the various error signals are obtained from the photo-detectors of the CD optical pick-up*

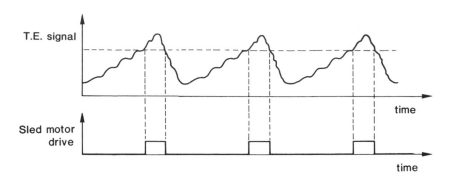

Figure 11.10 *Tracking error and sled motor drive signals within the CD player*

The Disc Motor Servo Circuit

In Chapter 2 we saw how each frame of information on the disc starts with a sync word. One of the functions of the sync words is to control the disc motor.

The sync word frequency is compared against a fixed frequency (derived from a crystal oscillator) in a phase comparator and the motor is driven according to any frequency or phase difference.

As the length of the tracks increases linearly from the inner (lead-in) track to the outer (lead-out) track, the number of frames per track increases in the same respect. This means that the frequency of the sync words also increases, which causes the motor speed to decrease, resulting in a constant linear velocity. The angular velocity typically decreases from 500 rpm (lead-in track) to 200 rpm (lead-out track).

As a practical example of the servo circuit ICs, the CXA1082 is one of the most used servo ICs. Figure 11.11 presents a block diagram of a CXA1082.

This IC has two defined parts: a digital control part and an analog servo control part.

The TTL-I2L registers are in/outputs to control the servo IC. In this way the servo IC will be instructed by the system control. Main inputs/outputs are to/from the system controller, which will trigger the execution of most operations. Besides these main instructions there are a number of dedicated inputs/outputs, for specific functions.

The analog part consists mainly of the focus and the tracking control lines. Focus error and tracking error are input from the RF part of the CD player.

These signals will be phase-compensated to correct any error. However, there are also a number of switches on these lines which are controlled by the sequencer. These switches enable manipulation of servo settings; in this way, gain can be set and power can be injected to drive the servo during specific operations; for example, if a track jump is requested, closing of TM3

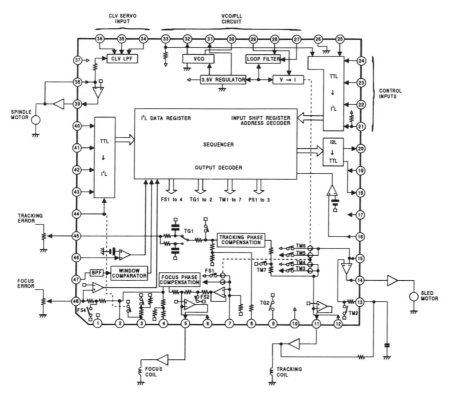

Figure 11.11 *CXA1082 block diagram*

and TM4 will inject the current to the tracking and sled coils for either direction.

The same circuits are used for initial control; thus, when the set is starting up it will try to focus the laser on the disc. In order to do this, the focus coil is driven up/down a number of times. At that moment the servo loop is disabled by FS4, and the up/down cycles are driven by FS1 and FS2.

Apart from these focus/tracking gain circuits this IC also comprises a VCO/PLL circuit and part of the spindle servo circuit.

Digital Servo

The latest development in servo circuitry is the use of digital servos. The benefits of digital servo circuits are listed below.

● Playability improved, as adjustments which used to be semi-fixed now become automatic and continuous.
● Adjustment-free, all adjustments are automatic, performed by the system itself.

● Use of DSP, this use of digital signal processors improves the possibilities and future features.

There are five servo circuits in a CD player: focus, tracking, sled, spindle and RF-PLL. Spindle and RF-PLL were digitized similar to drum motor/ capstan motor circuits used in VCR and DAT sets. Focus, tracking and sled servo were digitized using DSP and D/A technology. Figure 11.12 shows the main servo circuits.

Spindle and RF-PLL are using input from a reference cristal (RFCK) and from the RF retrieved from the disc. Focus, tracking and sled servo use the AD–DSP–DA setup shown here. Figure 11.13 shows a block diagram of digital servo.

● The A/D circuit will make 8-bit conversions of input signals at a rate of 88.2 kHz for focus and tracking error and at a rate of 345 Hz for sled error.
● The DSP circuit will carry out (digital) phase compensation; detection of specific signals (focus OK, defect and mirror) is also carried out based upon the input signals. Also, similar to the analog servo circuits, if specific actions need to be taken such as track jumps etc., this can be implemented easily by the DSP stages.

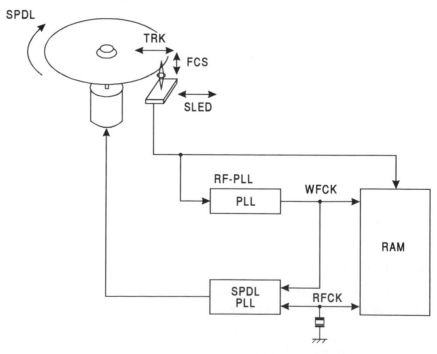

Figure 11.12 *Main servo circuits*

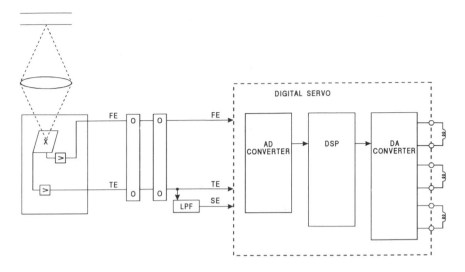

Figure 11.13 *Digital servo block diagram*

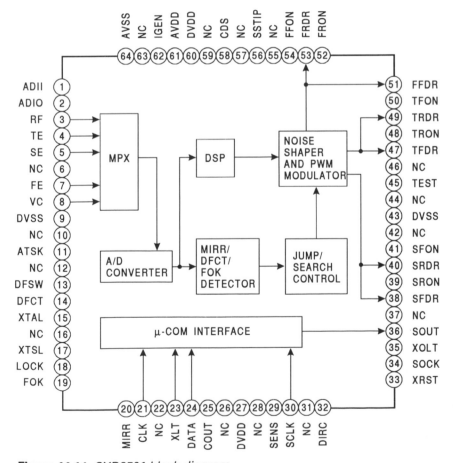

Figure 11.14 *CXD2501 block diagram*

● The D/A circuit will convert the error output signal to a 7-bit pulse width modulation (PWM) signal at 88.2 kHz for the focus and tracking coils and a 7-bit PWM signal at 345 Hz for the sled motor. The PWM signals will of course be fed to operational amplifiers where they are converted to a correct analog driving signal for coils and motor.

The CXD-2501 digital servo IC block diagram in Figure 11.14 shows the full integration of tracking, sled and focus servo into one IC.

12
Signal Processing

RF Amplification

The signal that is read from the compact disc and contains the data information is the RF signal, that consists of the sum of signals (A + B + C + D) from the main spot detector. At this stage, the signal is a weak current signal and requires current-to-voltage conversion, amplification and waveform shaping. The CDP-101 RF amplifier circuit is shown in Figure 12.1. IC402 is a current-to-voltage converter and amplifier stage, while IC403 is an offset amplifier, correcting the offset voltage of the RF signal and delivering the amplified RF signal.

Figure 12.2 shows the waveshaper circuit used in the CDP-101. Because

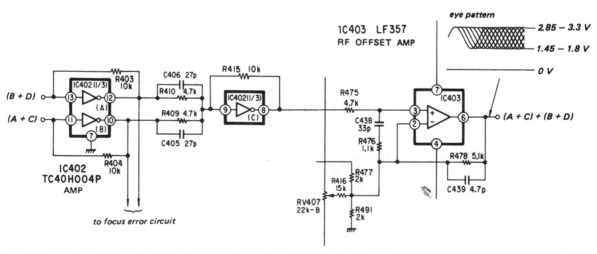

Figure 12.1 *RF amplification circuit of the CDP-101 compact disc player*

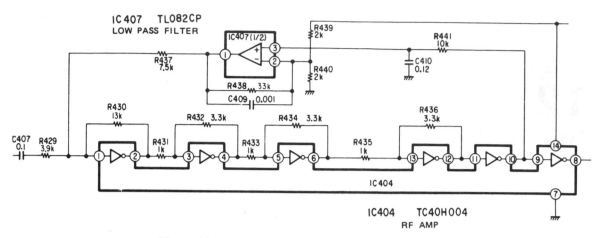

Figure 12.2 *Waveshaping circuit used in the CDP-101*

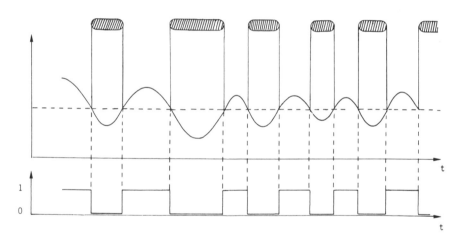

Figure 12.3 *Timing diagram of the RF signal before and after waveshaping*

the RF signal from the RF amplifier is a heterogeneous signal due to disc irregularities, the waveshaper circuit detects correct zero-cross points of the eye pattern and transforms the signal into a square wave signal. After waveshaping by IC404, the signal is integrated in the feedback loop through a low-pass filter circuit to obtain a DC voltage applied to the input, so as to obtain correct slicing of the eye pattern signal.

Figure 12.3 shows a timing diagram of the RF signal before and after waveform shaping.

Signal Decoding

The block diagram in Figure 12.4 represents the basic circuit blocks of a compact disc player. After waveshaping the RF signal is applied to a phase locked loop (PLL) circuit in order to detect the clock information from the signal which, in turn, synchronizes the data.

Also, the RF signal is applied to an EFM demodulator and associated error stages in order to obtain a demodulated signal. In the CDP-101 a single integrated circuit, the CX7933, performs EFM demodulation; a block diagram is shown in Figure 12.5.

This block diagram shows frame sync detection, 14-to-8 demodulation to a parallel 8-bit data output, subcode Q detection and generation by an internal counter and timing generator of the WFCK (write frame clock) and WRFQ (write request) synchronization signals. Figure 12.6 shows the EFM decoding algorithm (for comparison with the encoding scheme in figure 9.2).

CIRC decoding is performed by a single integrated circuit, the CX7935 (Figure 12.7) on the data stored in the RAM memory. A RAM control IC, the CX7934 (Figure 12.8) is used to control data manipulations between the RAM and the rest of the demodulation stage circuits.

The data are checked, corrected if necessary, and deinterleaved during readout. If non-correctable errors are found, a pointer for this data word is stored in memory and the circuit corrects the data by interpolation. Figure

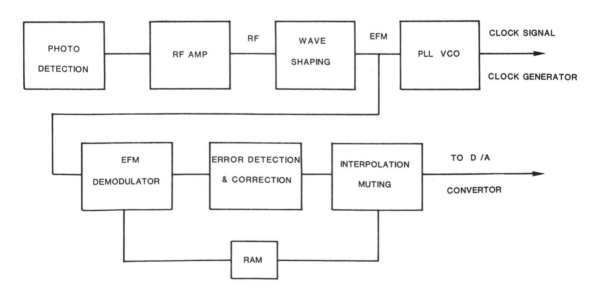

Figure 12.4 *Signal decoding within the compact disc player*

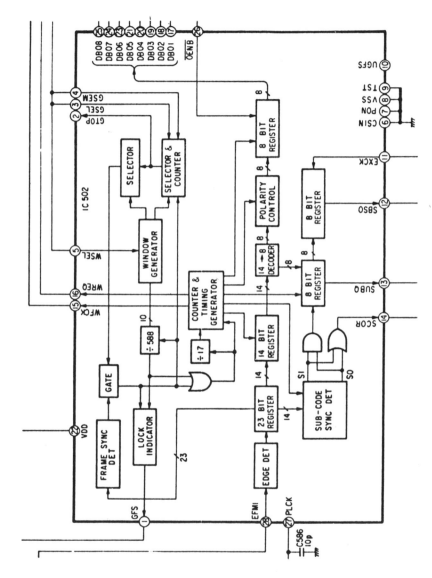

Figure 12.5 *Block diagram of the CX7933 integrated circuit: which performs EFM demodulation in the CDP-101*

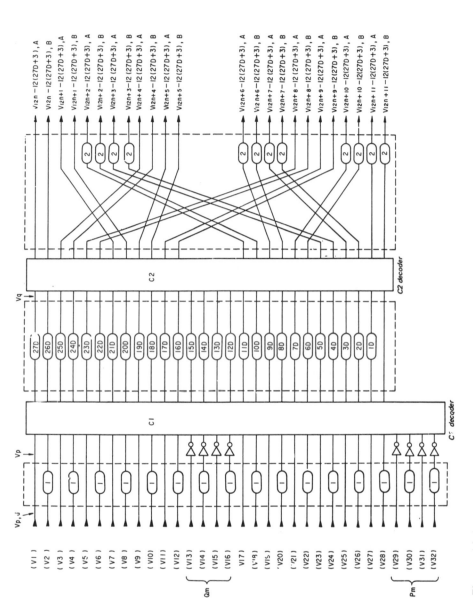

Figure 12.6 CIRC decoding algorithm

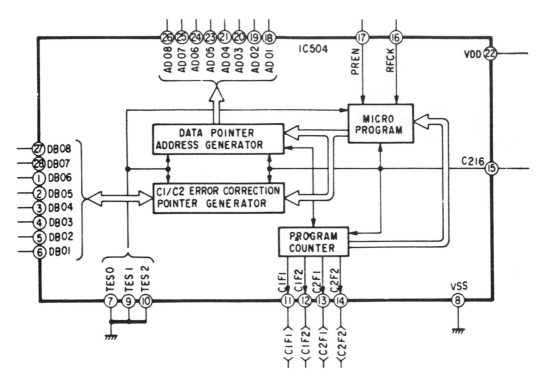

Figure 12.7 *Block diagram of the CX7935 CIRC decoder*

12.9 shows the complete signal decoding circuit as used in the CDP-101. In the latest Sony CD players, signal decoding is performed by a single integrated circuit, the CX23035, shown in Figure 12.10.

D/A Converter

The D/A converter follows the signal processing and decoding circuits. Figure 12.11 represents the CX20017 integrated circuit D/A converter as used in Sony CD players. The converter is formed around an integrating dual-slope A/D converter. Two counters, one for the eight most significant and one for the eight least significant bits, control the two constant current sources (which have a ratio: $I_o/i_o = 256$) used for charging the integrator capacitor. Conversion is controlled by the LRCK (left/right clock), BCLK (bit clock) and WCLK (word clock) signals.

Figure 12.12 represents the operating principle: where 16-bit data are loaded into the two 8-bit counters by a latch signal. With data in the counters, no carry signals exist and the current switches are closed. The integration capacitor C charges with a total current $I = I_1 + I_2$.

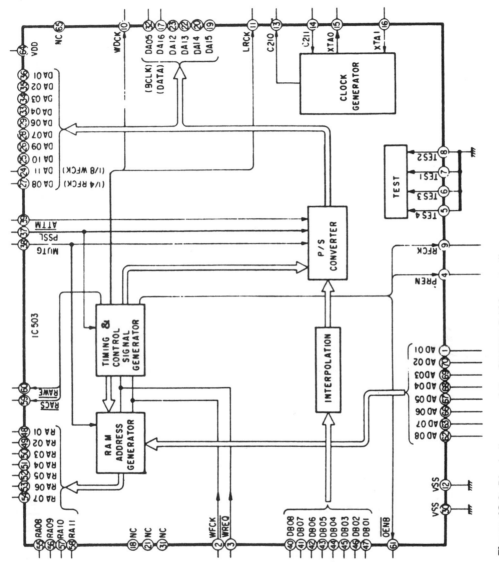

Figure 12.8 *Block diagram of the CX7934 RAM controller*

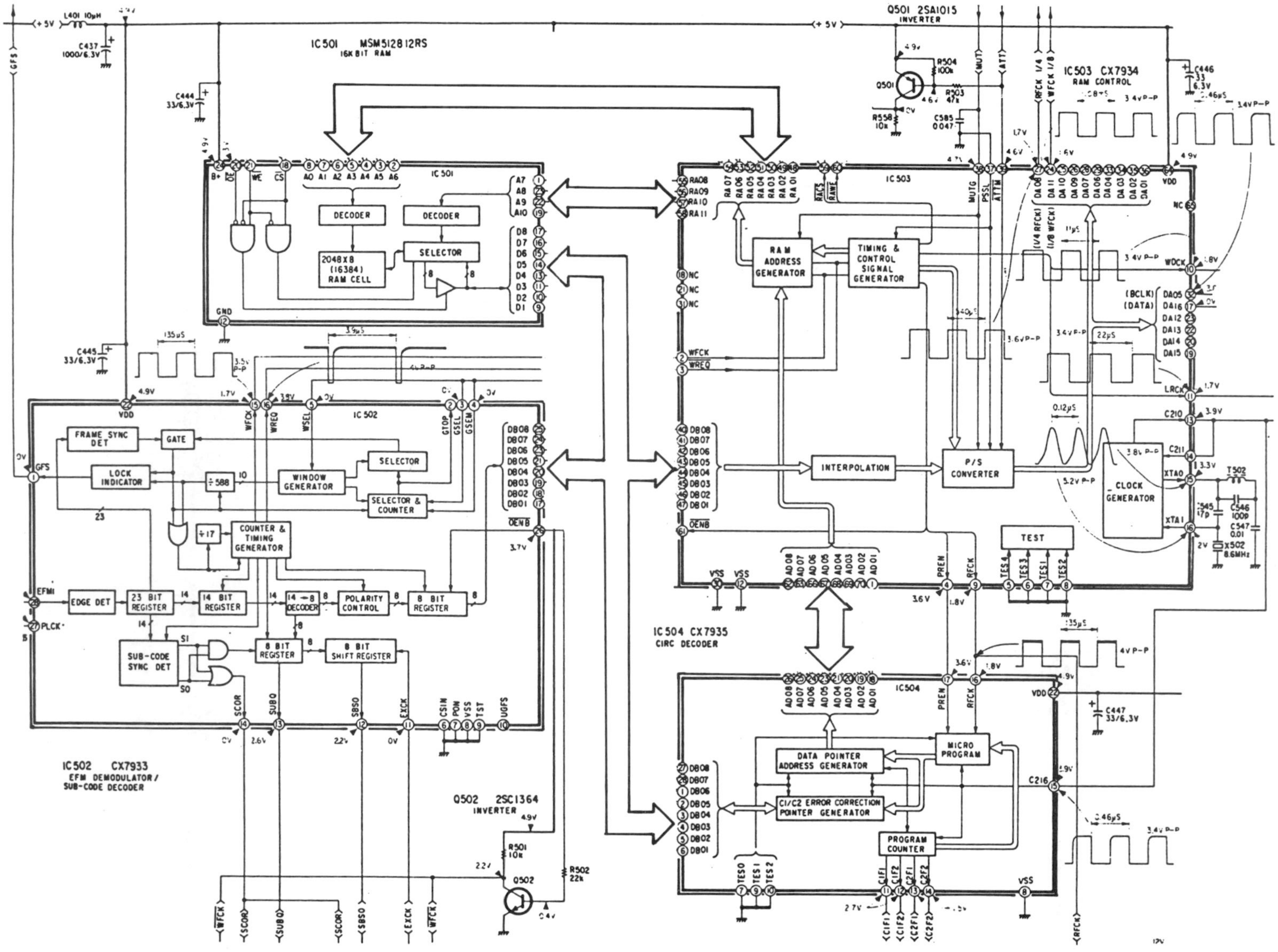

Figure 12.9 *Complete signal decoding circuit of the CDP-101*

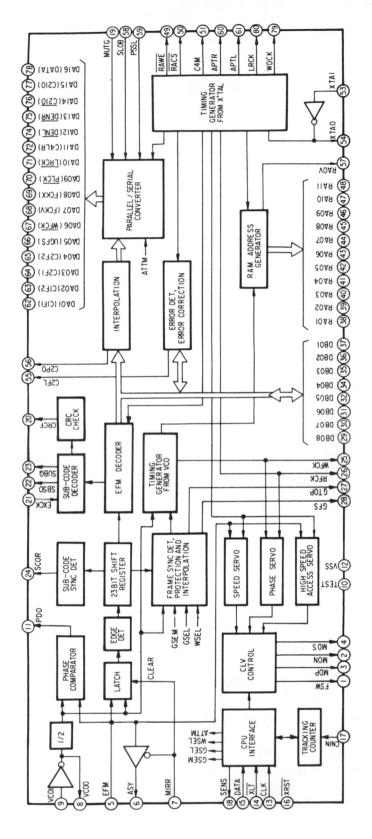

Figure 12.10 Block diagram of the CX23035 integrated circuit, which performs all signal decoding within the latest Sony CD players

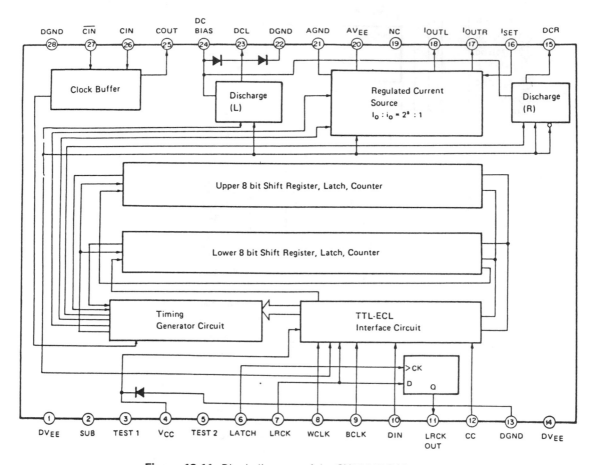

Figure 12.11 *Block diagram of the CX20017 D/A converter*

During conversion, each counter counts down to zero, whereupon the carry signals open the current switches, stopping further charging of the capacitor. The final charge across the capacitor, as an analog voltage, represents the 16-bit input.

Figure 12.13 shows a practical application of a D/A converter circuit in a CD player.

High Accuracy D/A Conversion

18-Bit digital filter/8-Times Oversampling

The CXD-1144 is a digital filter allowing the conversion of 16 bit samples into 18 bit samples with very high precision (Figure 12.14). The remaining ripple in the audible range is reduced to ± 0.00001 dB. The attenuation is 120 dB and the echo rejection is about 124 dB. Especially the reproduction of pulse-

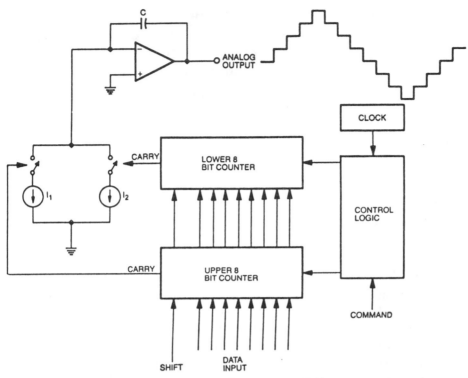

Figure 12.12 *Operating principle of the CX20017 D/A converter*

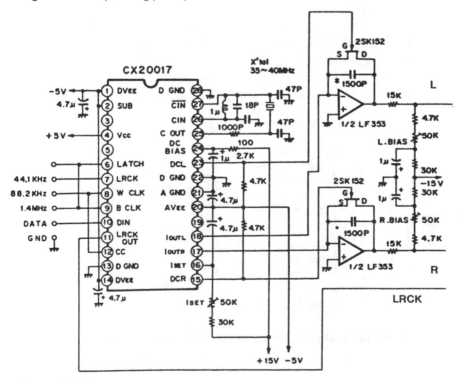

Figure 12.13

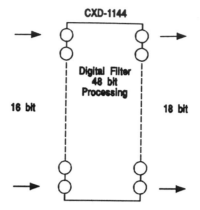

Figure 12.14 *CXD-1144 digital filter*

shaped tones, such as those of a keyboard or piano, is remarkably improved by enhancing the rising edges of the signal and the high echo rejection.

The digital filter is also used to create oversampling, calculating the intermediate values of the samples. An oversampling of 8 F_s or 352.8 kHz can be achieved. This increase in sampling rate gives an important reduction of quantization noise, which allows a more pure and analytic playback of music. Also a lower order LPF can be used, improving the group delay and linearity in the audio range.

In order to cope with 18 bit and such a high conversion rate great care must be taken in the designing of the D/A converter. To reduce the load imposed on the D/A converter Sony developed an 'overlapped staggered D/A conversion system' (Figure 12.15). The basic idea is to use a digital filter circuit at 8 F_s output, combined with two D/A converters for each channel.

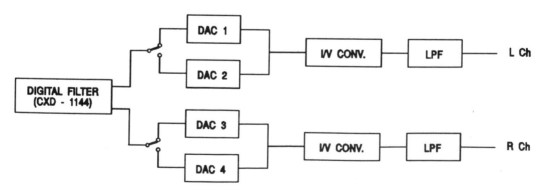

Figure 12.15 *Overlapped staggered D/A conversion system*

The conversion rate is 4 F_s, so the digital filter output is at $2 \times 4 = 8 \, F_s$ oversampling. The even and odd samples for each channel are applied to separate D/A converters. By adding the output of both DAC, corresponding to the formula: $(L1 + L2)/2$, a staircase signal is obtained which corresponds with an 8 F_s oversampled output signal. Since the outputs of the two DACs are added continuously a maximum improvement of 3 dB is realized in quantization noise with reduced distortion. The output current is also doubled, improving the S/N ratio of the analog noise by a maximum of 6 dB.

High Density Linear Converter

Figure 12.16 shows a block diagram of a single bit pulse D/A converter. The digital filter, CXD-1244, uses an internal 45 bit accumulator to perform accurate oversampling needed in the single bit converter. The pulse D/A converter combines a third-order noise shaper and a PLM (Pulse Length Modulation) converter to produce a train of pulses at its output. By using a low-order LPF the analog signal is obtained. A digital sync circuit is inserted between the DAC and the digital filter to prevent jitter of the digital signal.

Compared with conventional D/A converters the high density linear converter provides highly accurate D/A conversion with improved dynamic range and extremely low harmonic distortion.

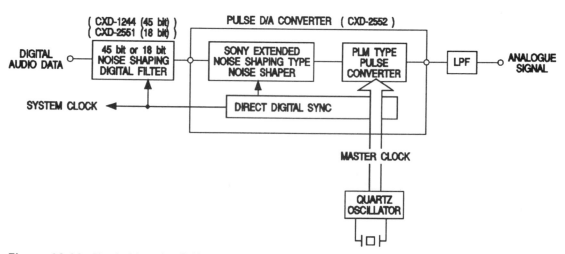

Figure 12.16 *Single bit pulse D/A converter*

PART THREE

Digital Audio Recording Systems

13
Outline

The advantages of digital techniques were first realized in the field of magnetic recording. Analog magnetic recording creates significant deterioration of the original sound: using analog techniques, for example, it is difficult to obtain a flat frequency response at all signal levels. Further, signal-to-noise ratio is limited to some 70 dB, the sound is deteriorated by speed variations of the recorder mechanism, there exist crosstalk and print-through problems, and any additional copying deteriorates the characteristics even further. In addition to this, to keep the equipment within close specifications, as required in a professional environment, frequent and costly realignment and maintenance are required.

Digital magnetic recording, on the other hand, virtually solves all of these drawbacks. Recording of digital data, however, presents some specific problems:

- required bandwidth is increased dramatically compared to the original signal
- specific codes must be used for recording (in contrast to the simple data codes mentioned before)
- error-correction data must be recorded
- synchronization of the recorded data stream is necessary to allow for reconstruction of the recorded words
- in contrast to analog recordings, editing is very complicated and requires complex circuits. For tape-cut editing, common practice in the analog recording field, a very strong error-correction scheme together with interleaving are needed. Even then, very careful handling is a must; for instance, the tape cannot be touched with bare fingers.

Several different techniques have been developed, outlined in the following chapters.

14
PCM Adapters According to EIAJ

PCM adapters or processors convert the audio information into a pseudo-video signal for subsequent recording on a video recorder. The EIAJ standard was basically established as a format for consumer applications. The basic specifications of the EIAJ standard are listed in Table 14.1.

Sony processors such as the PCM-F1, PCM-701ES and PCM-501ES also have a 16-bit recording mode and have become very popular in the professional recording field. 16-bit mode recordings can be played back on either 16-bit or 14-bit machines. The PCM-F1 for instance has the following specification:

Frequency response:	$10 - 20,000\,\text{Hz} \pm 0.5\,\text{dB}$
Dynamic range:	>86 dB (14-bit mode)
	>90 dB (16-bit mode)
Harmonic distortion:	<0.007% (14-bit mode)
	<0.005% (16-bit mode)
Wow and flutter:	Unmeasurable

The EIAJ format's error-correction system is called **b-adjacent coding**, a system which adds two error-correction words, called P and Q, to six data words. In the 16-bit mode the Q word is not used, and the space available is taken by the extra bits of the signal information. Therefore, the error-correction capability of the 16-bit format is slightly inferior to the 14-bit format. In practice, however, as a good concealment system takes care of possible uncorrected errors to the same extent as in the 14-bit mode, this is of no real consequence. A comparison of error-correction capabilities of EIAJ format processors and the PCM-F1 16-bit mode is given in Table 14.2. Time corresponding to a horizontal video line is given the symbol, H.

Table 14.1 *EIAJ specification for digital audio tape recorders*

Item	Specification
Number of channels	2 (CH-1 = left, CH-2 = right)
Number of bits	14 bits/word
Quantization	linear
Digital code	2's complement
Modulation	non-return to zero
Sampling frequency	44.056 kHz (NTSC)/44.1 kHz (PAL)
Bit transmission rate	2.634×10^6 bit s^{-1} (NTSC)/ 2.625×10^6 bit s^{-1} (PAL)
Video signal	NTSC standard/PAL standard

Table 14.2 *Error-correction capabilities of EIAJ and PCM-F1 format audio digital tape recorders*

	Error-correction word	Error-correction capability	Range of compensation (concealment)	Muting condition
EIAJ 14-bit format	P, Q	Burst error less than 32H	32H–95H	Burst error more than 95H
PCM-F1 16-bit format	P only	Burst error less than 16H	16H–95H	Burst error more than 95H

A/D Conversion

The EIAJ format specifies a 14-bit linear conversion system which allows a theoretical dynamic range of 86 dB. Pre-emphasis can be applied, with turn-over frequencies of 3.18 kHz and 10.6 kHz (i.e., the same as in the CD format). Characteristics of the EIAJ format pre-emphasis and de-emphasis are shown in Figure 14.1.

When a recording is made with pre-emphasis switched in, a control signal on the tape records this fact. During playback, switching of the de-emphasis circuit then occurs automatically.

The control signal also, among other things, contains a 'copy protect' bit which, when set, prevents copies being made of a recording.

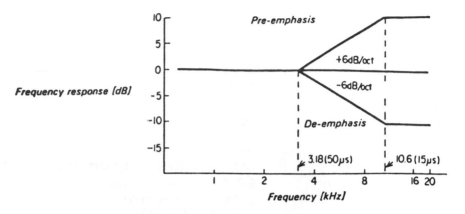

Figure 14.1 *EIAJ format pre-emphasis and de-emphasis characteristics*

Encoding System

The processor samples the left (L) and right (R) audio channels alternately and, after undergoing A/D conversion, data are fed to six input terminals of the encoder, in the sequence shown in Figure 14.2.

Two error-correction words P and Q are generated with an exclusive-or operation and a matrix operation. Then, the input data, together with the derived P and Q words, are **interleaved** by using time delays formed by RAM memory stages. The delays are in multiples of a horizontal video line time H, and one word is encoded through each input terminal of the encoder in time H.

Starting at time 0H, consider input word L0. As this word undergoes no delay, it appears at time 0H at the output A. Input word R0, on the other hand, will not appear at the output until time 16H. Hence during the time 0H to 15H, data line A comprises the sequence of words:

L0 L3 L6 L9 L12 L15 ... L45

while nothing comes out of the other lines.

At time 16H, line B starts to output data. Hence the data at lines A and B is:

Line A   L48
Line B   R0

Still, nothing comes out from the other lines.

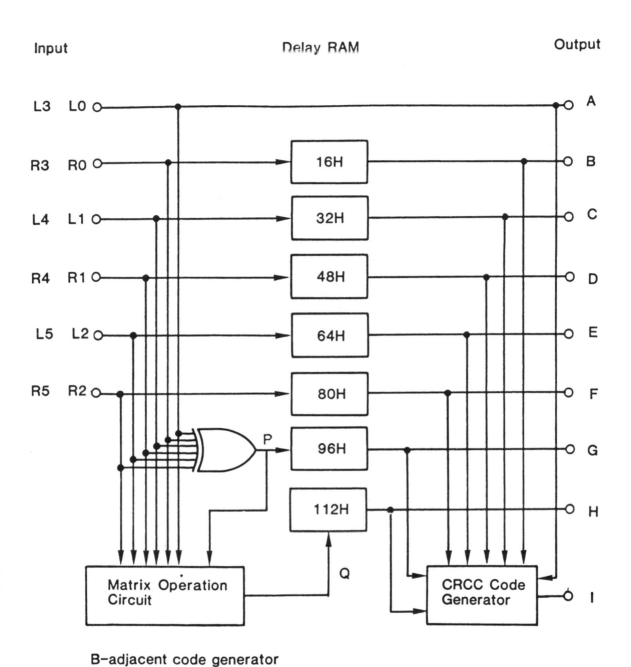

Input   Delay RAM   Output

L3   L0 ── A

R3   R0 ──────────────── 16H ─────────────────── B

L4   L1 ──────────────── 32H ─────────────────── C

R4   R1 ──────────────── 48H ─────────────────── D

L5   L2 ──────────────── 64H ─────────────────── E

R5   R2 ──────────────── 80H ─────────────────── F

P ──────────────── 96H ─────────────────── G

112H ─────────────────── H

Matrix Operation Circuit   Q   CRCC Code Generator   I

B-adjacent code generator

Figure 14.2 *Encoding system for a digital audio tape recorder*

Similarly, at time 32H, line C starts to output data and we have the following:

Line A   L96
Line B   R48
Line C   L1

In general, for nH, data outputs at lines A, B and C are as follows:

Line A L_3n
Line B $R_3(n-16)$
Line C $L_3(n-32)+1$

As the process continues, the output data at lines A–H are as shown in Figure 14.3. Note that after this operation, R0 instead of L0 is paired with L48. Similarly, R1, R2 are paired with L49 and L50 respectively.

When data are recorded on the magnetic tape in this way, errors due to dropouts are effectively avoided. Even when a large dropout occurs and

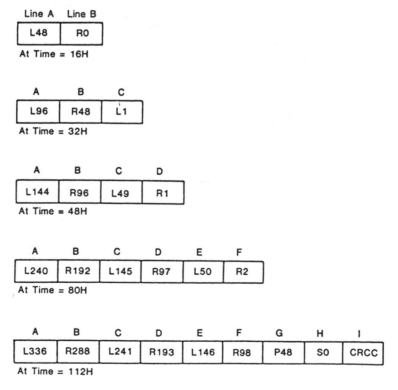

Figure 14.3 *Output data from the encoder, at various times*

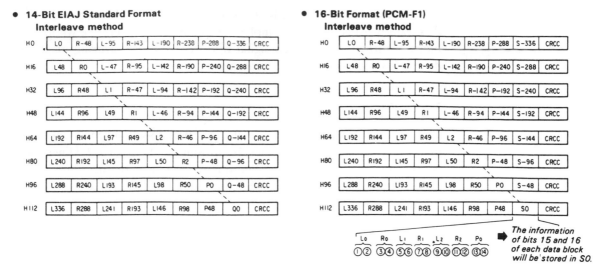

Figure 14.4 *Interleaving formats of EIAJ and PCM-F1 systems*

significant numbers of data words are lost, the spread of words over a wide area due to interleaving means that many words will still be available. Error correction can be applied in most cases of dropout.

Figure 14.4 shows the interleave formats of the EIAJ 14-bit system and the PCM-F1 16-bit system. In the 16-bit format, Q error correction words are not used. Instead, S words, which contain bits 15 and 16 of six data words and their corresponding P word, are recorded.

A significant amount of redundancy is incorporated into the data signal to ensure error detection and correction. As a percentage, redundancy is calculated as the ratio of the number of error-correction bits to the number of audio bits plus error-correction bits:

a) **14-bit mode**

There are $6 \times 14 = 84$ audio data bits and $(2 \times 14) + 16 = 44$ error correction and detection bits, so redundancy R, is:

$$\frac{44}{44 + 84} = \frac{44}{128} = 34.4\%$$

b) **16-bit mode**

There are $6 \times 16 = 96$ data bits and $14 + 16 = 30$ error correction and detection bits, so redundancy R, is:

$$\frac{30}{30 + 96} = \frac{30}{126} = 23.8\%$$

After interleaving, the CRCC (cyclic redundancy check code) error-detection word is added to complete one data block. Then, the whole data block is turned into a video signal prior to recording. During playback, **de-interleaving** is applied to recreate the original data.

Video Format

The encoded data signals are modulated on a pseudo-video signal. One frame of a PAL video signal has 625 lines, comprising two fields of 312.5 lines each. NTSC video signals have a 525-line frame, with fields of 262.5 lines. Each field is preceded by vertical synchronization pulses, while each line is preceded by a horizontal sync pulse.

Format in one field

At field level, the data are coded as shown in Figure 14.5a, where the data lines are preceded with one line that contains a control word C.

This control word is illustrated in Figure 14.5b, and comprises:

a) cueing signal: a repetition of fourteen '1100' signals for detection purposes
b) ID: content identification, normally not used
c) address: not used
d) control: only the last four bits have been defined, as listed in Table 14.3
e) CRCC: a 16-bit error word on all the preceding data

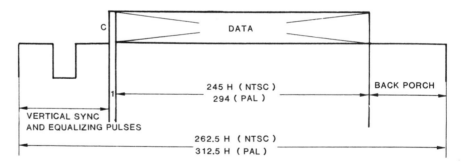

CUEING SIGNAL	ID	ADDRESS	CTL	CRCC
56	14	28	14	16

128 bits

Figure 14.5 *Video format of encoded data (a) in each field (b) the control word C*

Table 14.3 *Allocated control bits of the control word C*

BIT	MEANING	CODE
11	Copy-prohibiting	0 = none
12	P-correction identification	0 = present
13	Q-correction identification	0 = present
14	Pre-emphasis identification	0 = present

Format in one line

One horizontal line, shown in Figure 14.6, comprises 168 bits, 128 bits of which are data bits. This means that the PAL data rate is:

$$\frac{168}{64 \ \mu s} = 2.625 \times 10^6 \ \text{bit s}^{-1}$$

and data rate in NTSC is:

$$\frac{168}{63.6 \ \mu s} = 2.643 \times 10^6 \ \text{bit s}^{-1}$$

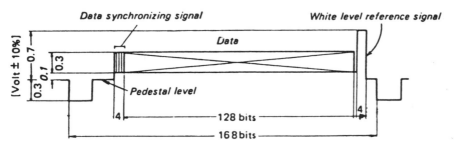

Figure 14.6 *Video format as number of bits in a line*

Basic Circuitry of a PCM Processor

A block diagram of a PCM processor, such as the Sony PCM-F1, is given in Figure 14.7, and comprises:

● a recording circuit (A), which receives the analog audio input signals and, after processing, outputs a pseudo-video signal for recording using a standard video recorder
● a playback circuit (B), which receives the played back video signal from the video recorder and outputs, after processing, the reconstructed analog audio signal
● a support circuit (C), which provides the necessary timing and control signals for both processing circuits.

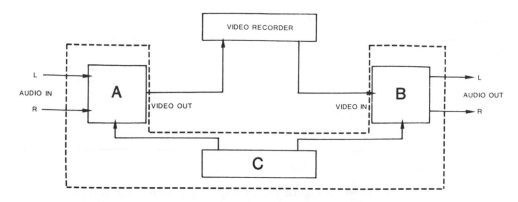

Figure 14.7 *Block diagram of a PCM processor system*

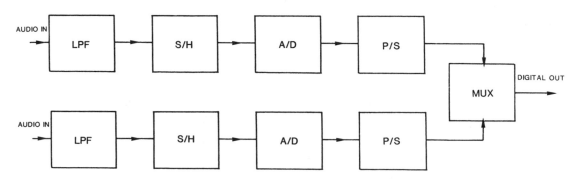

Figure 14.8 *Input stage of the recording circuit*

The Recording Circuit

Input circuit

A block diagram of the input stage of the recording circuit is shown in Figure 14.8, where the analog audio input signals are fed through an anti-aliasing, low-pass filter (LPF) which limits the bandwidth of the signal to half the sampling frequency.

These audio signals, analog and continuous in time, are transformed in the sample and hold (S/H) into analog signals which are discrete in time.

The analog-to-digital converter (A/D) converts the analog signals into digital ones (discrete in value *and* in time).

The parallel output signals are converted into a serial bit stream by a parallel to serial (P/S) circuit and both channels are combined in a multiplexer (MUX) which outputs a serial digital data bit stream.

The signal processor

Figure 14.9 shows a block diagram of the signal processing stage of the recording circuit.

An error-protection circuit calculates an error-detection and correction word based upon the information contained in the input signal. Error-detection and corrections words are inserted at regular intervals, into the signal, giving a measure of protection against single-bit errors.

To protect the information against burst errors (caused by, say, tape drop-out), the signal is interleaved. This effectively converts burst errors into single-bit errors which can be corrected.

A time base corrector (TBC) compresses the signal in time, so that it matches the characteristics of a standard video signal. Frequently, a RAM arranged as a first-in-first-out (FIFO) buffer is used as a TBC.

A digital-to-video converter (DVC) finally adds the necessary horizontal and vertical synchronization signals and outputs a pseudo-video signal which can be recorded using a video recorder.

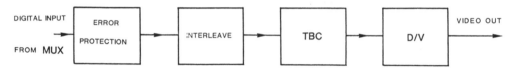

Figure 14.9 *Signal processing stage of the recording circuit*

The Playback Circuit

The signal processor

A block diagram of the signal processing stage of the playback circuit is shown in Figure 14.10, where a video-to-digital converter (V/D) isolates the digital data from the horizontal and vertical synchronization signals.

A time base corrector (TBC) like that of the recording circuit; a FIFO buffer, receives the input signals. The horizontal synchronization signal is used as a clock signal for reading in the digital data while the processor's

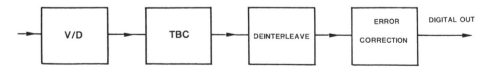

Figure 14.10 *Signal processing stage of the playback circuit*

system clock (crystal controlled) controls data output. The TBC serves two main purposes:

1 removal of possible remaining jitter on the input signals, caused by unstable video recorder playback
2 time expansion of the video signal into a continuous data stream.

A de-interleave circuit rearranges the data bits into their correct sequence, at the same time converting possible burst errors into single-bit errors.

An error-correction circuit uses the redundant information, added to the signal in the error-protection circuit as recording, to detect and correct single-bit errors. A burst error exceeding the correction capability of the system will be concealed (by interpolation, previous word hold or simple muting means).

Output of this stage is a digital serial data stream, ready to be converted into analog.

The output stage

The output stage of the playback circuit is shown in a block diagram in Figure 14.11.

A demultiplexer (DEMUX) separates the digital input signal into left and right channels. A serial to parallel converter (S/P) combines the serial data bits into parallel data words which are then converted into analog signals by a digital-to-analog converter (D/A). Output of the D/A converter is a pulse amplitude modulated (PAM) signal.

An aperture control circuit (AC), a sample and hold, removes glitches from the signal and corrects frequency response.

A low-pass filter (LPF) performs final reconstruction by removing images from the audio signal (for this reason, this type of filter is sometimes referred to as an **anti-imaging filter**).

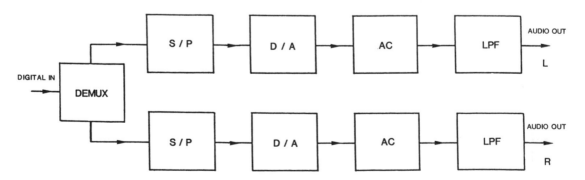

Figure 14.11 *Output stage of the playback circuit*

15
PCM-1600/1610 Format

In professional recording studios, the PCM-1600/1610 format appears to be the de facto standard for compact disc mastering, as virtually all compact discs are mastered on either PCM-1610 or PCM-1630 processors. The format was established as a two-channel studio recording standard with extremely strong error-correction capabilities to enable electronic editing.

Basically, a professional PCM processor, such as the PCM-1610, has the same principles of operation as an EIAJ processor. There are, however, differences in the error-correction techniques used. Table 15.1 summarizes the main points of the PCM-1600/1610 format.

Encoding Scheme

The error-correction system adopted in the PCM-1600/1610 format was developed by Sony and is called 'crossword code'. In the crossword system, error-correction and detection words are added as shown in Figure 15.1,

Table 15.1 *PCM-1600/1610 format specification*

Item	Specification
Number of channels	2(CH-1 = left, CH-2 = right)
Number of bits	16 bits/word (per channel)
Quantization	linear
Digital code	2's complement
Sampling frequency	44.056 kHz or 44.1 kHz
Bit transmission rate	3.5795 or 3.5831 $\times$ 10^6 bit s^{-1}
Video signal	NTSC standard

R1	L2	R3	C1
P1	P2	P3	C3
L1	R2	L3	C2

Figure 15.1 *Illustration of the crossword code error-detection system*

where a block of six audio data words (three left channel words and three right channel words) is linked with three parity words and three CRCC words.

The parity words are generated by an exclusive-OR function performed on the left and right channel words, where

$$Pn = Ln + Rn$$

The CRCC words are generated by the polynomial:

$$X^{16} + X^{12} + X^5 + 1$$

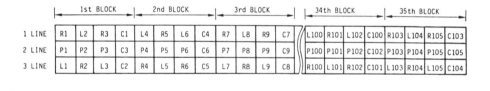

Figure 15.2 *Interleaving data prior to recording*

As a basic block consists of six audio words, and six error-correction and detection words, redundancy is 50%.

Figure 15.2 shows how data are then interleaved. A total of 420 interleaved words, i.e., 105 L-data words, 105 R-data words, 105 P-data words and 105 CRCC words make one interleave, recorded as 35 horizontal lines. A complete video field therefore stores some seven interleaves (i.e., 245 data lines).

Video Format

The format of one video line is shown in Figure 15.3. This corresponds to the first 12 words of line 1 of Figure 15.2. As a complete interleave takes 35 video

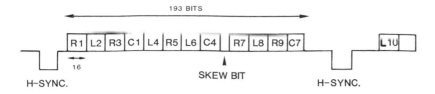

Figure 15.3 *Format of a single video line*

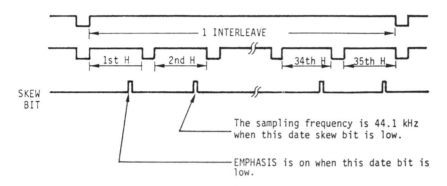

Figure 15.4 *Showing a complete video interleave of data, along with skew bits*

Table 15.2 *Settings and meanings of the two skew bits*

Skew bit (1st line)	Skew bit (2nd line)	Sampling frequency (kHz)	Emphasis ON/OFF
0	0	44.1	ON
0	1	44.056	ON
1	0	44.1	OFF
1	1	44.056	OFF

lines, the first 12 words of line 2 of Figure 15.2 will be written in the 36th line following that shown in Figure 15.3, and so on.

The 129th bit of each horizontal line is a **skew** bit, used to contain data regarding sampling frequency and emphasis. The skew bit of the first and second horizontal lines is set as shown in Table 15.2; the skew bits of the other horizontal lines are always 1. Use of the skew bit is illustrated in Figure 15.4, where signals corresponding to a complete interleave are shown.

16
Video 8 PCM-Format

In 1985, a new consumer video format was launched, called Video 8. Sony expects this format to replace gradually the older Betamax and VHS consumer video formats. The video 8 format uses a much smaller cassette than older video formats, enabling construction of very small video recorders. Almost all major manufacturers in the field of consumer electronics are supporting this new format.

Audio information can be recorded on Video 8 recorders as either an FM signal, along with the picture, or as a PCM signal written in a section of the tape where no picture information is recorded. In some recorders, however, PCM data can be recorded in the video area, instead of the picture signal. By doing this, six channels of high-quality audio can be recorded on a tape.

The specification of the Video 8 PCM standard is summarized in Table 16.1.

A/D–D/A Conversion

As only 8 bits per channel are used, audio characteristics would be poor if special measures were not taken. These measures include:

- audio compression and expansion for noise-reduction purposes
- 10-bit sampling
- non-linear quantization by 10-bit to 8-bit compression and expansion

and are illustrated, in a block diagram, in Figure 16.1. Characteristic of the noise-reduction (NR) system is shown in Figure 16.2, while Figure 16.3 shows the characteristic of the non-linear encoder. The upper limit of the frequency response of the system is limited to a maximum of 15,625 Hz, i.e., half of the sampling frequency of 31,250 Hz.

Photo 16.1 *Video 8 cassette*

Table 16.1 *Specification of Video 8 format*

Item	Specification
Number of channels	3 (CH-1 = left, CH-2 = right)
Number of bits	8 bits/word
Quantization	linear
Digital code	2's complement
Modulation	bi-phase (FSK) 2.9 MHz, 5.8 MHz
Sampling frequency	31,250 kHz (PAL)/
	31,468.53 Hz (NTSC)
Bit transmission rate	5.8×10^6 bit s^{-1}
Video signal	PAL/NTSC
Number of PCM tracks	6

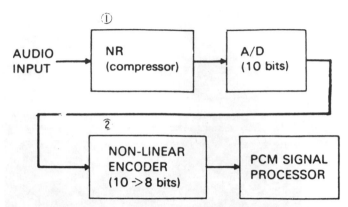

Figure 16.1 *Block diagram of Video 8 signal processing*

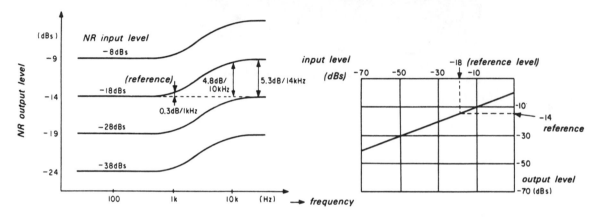

Figure 16.2 *Noise reduction system characteristic*

With these measures, typical audio characteristics of Video 8 PCM audio recordings are:

● frequency response: 20–15,000 Hz
● dynamic range: more than 88 dB
● sampling frequency: 31.5 kHz
● quantization: 8-bit non-linear
● wow and flutter: less than 0.005%

Description of the Format

One track on the section of tape used to record PCM audio information holds 157 blocks of data for a PAL machine, 132 blocks for an NTSC machine. Each block contains eight 8-bit data words, two 8-bit parity words (P and Q), one 16-bit error-detection word, one 8-bit address word, and three sync bits.

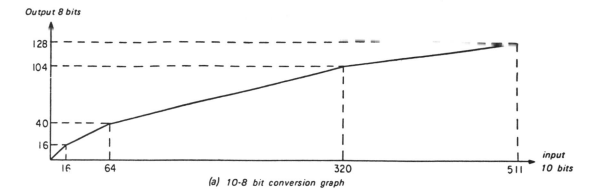

(a) 10-8 bit conversion graph

Y=X	(0≦X<16)
Y=|X/2|+8	(16≦X<64)
Y=|X/4|+24	(64≦X<320)
Y=|X/8|+64	(320≦X≦511)

X: Input absolute value

Y: Output absolute value

(b) Conversion algorithm

Encode law	Input data	Output data
10bits→10bits	0—15	0—15
10bits→9bits	16—63	16—39
10bits→8bits	64—319	40—103
10bits→7bits	320—511	104—127

(c)

Figure 16.3 *Non-linear encoder characteristic*

So one block comprises 107 bits and each track comprises 16,799 bits in PAL mode, 14,017 bits in NTSC mode.

Error-correction and detection words are added as shown in Figure 16.4.

The error-correction code adopted for Video 8 PCM is a modified cross-interleaved code (MCIC) in which the code is composed of blocks which are related to the video fields. The version used is called **improved MCIC**, in which ICIC, the initial value, necessary for parity calculation, can be any value and has numerous applications, such as identification words.

As eight audio data words are combined with two parity words, the Video 8 system is often called an 8w-2p coding system. A CRCC word is also added as an error detector.

In encoding, the sequence can be expressed as follows:

P – parity

$$P(n) = Q(n + D) + \sum_{i = 2}^{8} W_{i-2}(n + iD)$$

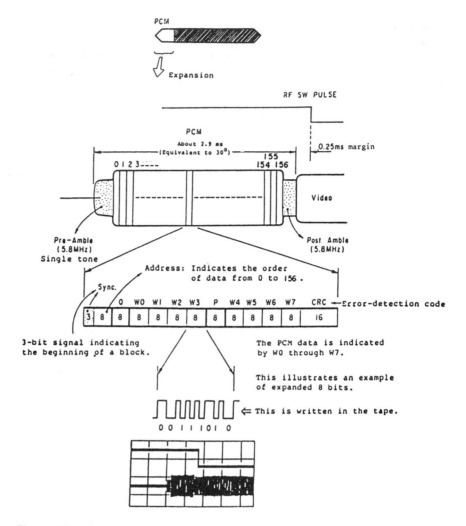

Figure 16.4 *Error-correction encoding prior to recording*

Q – parity sequence

$$Q(n + D) = P(n + d) + \sum_{i=2}^{8} W_{i-2}[n + i(D - d) + d]$$

where n is the block number ($0 < n < 157$ for PAL recordings, and $0 < n < 132$ for NTSC), D is the delay of the P parity sequence which converts a burst error into random errors (17 for PAL, 1514 for NTSC) and d is the Q parity sequence delay behind the P parity sequence (3 for PAL and 32 for NTSC).

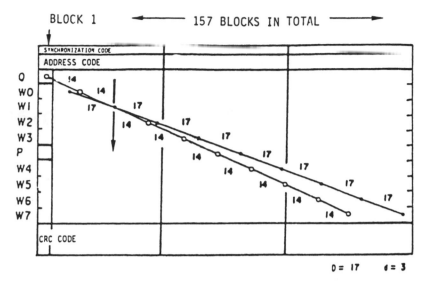

Figure 16.5 *Data interleaving in Video 8 PCM*

The error-detection code is a 16-bit CRCC and its polynomial is given by:

$$g(X) = X^{16} + X^{12} + X^5 + 1$$

In decoding, the pointer method is used, which corrects an erroneous word using a pointer flag.

The redundancy of the Video 8 format is as follows:

there are $8 \times 8 = 64$ audio data bits and,

$(2 \times 8) + 16 \qquad = 32$ error-correction and detection bits, plus

$1 \times 8 \qquad\qquad = 8$ address bits

so, redundancy R, is:

$$\frac{32 + 8}{32 + 8 + 64} = 38.5\%$$

Words are interleaved onto the PCM section of tape as shown in Figure 16.5.

17
Digital Audio Tape (DAT) Format

Although digital audio processors have been developed and used for many years, using conventional video recorders to store high-quality audio information, it is inevitable that some form of tape mechanism be required to do the job in a more compact way. Two main formats have been specified.

The first format, known as **rotary head, digital audio tape** (R-DAT), is based on the same rotary head principle as a video recorder, and so has the same limitations in portability. The second format, known as **stationary head, digital audio tape** (S-DAT), is currently in development and will use a stationary head technique mechanically similar to analog audio recorders.

Photo 17.1 *DAT mechanism*

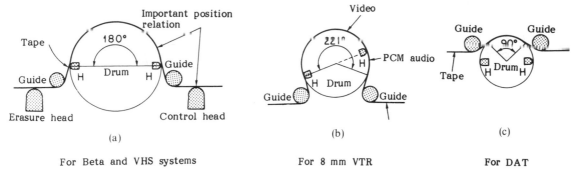

Figure 17.1

R-DAT

One important difference between standard video recorder and R-DAT techniques is that in a video recorder the recorded signal is continuous; two heads on the drum make contact with the tape for 180° each (i.e., the system is said to have a 180° wrap angle, as shown in Figure 17.1a), or 221° each (a 221° wrap angle, as in Figure 17.1b). In the R-DAT system, where the digital audio signal is time-compressed meaning that the heads only need to make contact with the tape for a smaller proportion of the time (actually 50%), a smaller wrap angle may be used (90° – as shown in Figure 17.1c).

This means only a short length of tape is in contact with the drum at any one time. Tape damage is consequently reduced, and only a low tape tension is necessary with resultant increase in head life.

The R-DAT standard specifies three sampling frequencies:

- 48 kHz; this frequency is mandatory and is used for recording and playback.
- 44.1 kHz; this frequency, which is the same as for CD, is used for playback of pre-recorded tapes only.
- 32 kHz; this frequency is optional and three modes are provided.
- 32 kHz has been selected as it corresponds with the broadcast standard.

Quantization:

- A 16-bit linear quantization is the standard for all three sampling rates.
- A 12-bit non-linear quantization is provided for special applications such as long play mode at reduced drum speed, 1000 rpm (–mode III) and U-channel applications.

Figure 17.2 shows a simplified R-DAT track pattern.

Table 17.1

Item	DAT (REC/PB mode)				Pre-recorded tape (PB only)	
Mode	Standard	Option 1	Option 2	Option 3	Normal track	Wide track
Channel number (CH)	2	2	2	4	2	2
Sampling frequency (kHz)	48	32	32	32	44.1	
Quantization bit number (bit)	16 (Linear)	16 (Linear)	12 (Nonlinear)	12 (Nonlinear)	16 (Linear)	16 (Linear)
Linear recording density (Kbit in^{-1})	61.0	61.0	61.0	61.0	61.0	61.1
Surface recording density (Mbit in^{-2})	114	114	114	114	114	76
Transmission rate (Mbit s^{-1})	2.46	2.46	1.23	2.46	2.46	2.46
Subcode capacity (Kbit s^{-1})	273.1	273.1	136.5	273.1	273.1	273.1
Modulation system	8–10 conversion					
Correction system	Double Reed-Solomon code					
Tracking system	Area sharing ATF					
Cassette size (mm)	73 × 54 × 10.5					
Recording time (min)	120	120	240	120	120	80
Tape width (mm)	3.81					
Tape type	Metal powder				Oxide tape	
Tape thickness (μm)	13 ± 1 μm					
Tape speed (mm s^{-1})	8.15	8.15	4.075	8.15	8.15	12.225
Track pitch (μm)	13.591	13.591	13.591	13.591	13.591	20.41
Track angle	6°22'59.5"					6°23'29.4"
Standard drum specifications	30 mm diameter 90° wrap					
Drum rotations (rpm)	2000	2000	1000	2000	2000	2000
Relative speed m s^{-1}	3.133	3.133	1.567	3.133	3.133	3.129
Head azimuth angle	±20°					

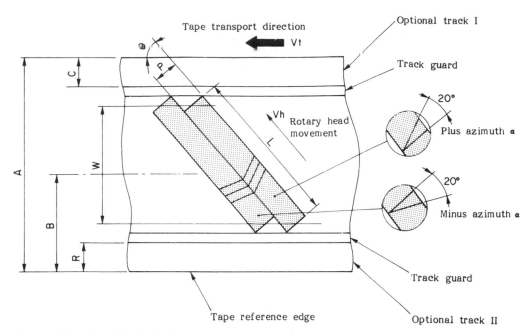

Figure 17.2 *Simplified R-DAT track pattern*

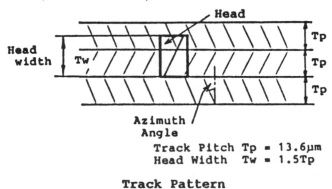

Track Pitch Tp = 13.6μm
Head Width Tw = 1.5Tp

Track Pattern

Figure 17.3 *Overwrite recording is used to ensure each track is as narrow as possible and no guard-band is required*

The standard track width is $13.591\,\mu m$, the track length is $23.5\,mm$, the linear tape speed is $8.1\,mm\,s^{-1}$. The tape speed of the analog compact cassette (TM) is $47.6\,mm\,s^{-1}$. This results in a packing density of $114\,Mbits\,s^{-1}\,m^{-2}$. See Table 17.1.

The R-DAT format specifies a track width of only $13.6\,\mu m$, but the head width is about 1.5 times this value, around $20\,\mu m$. A procedure known as **overwrite recording** is used, where one head partially records over the track recorded by the previous head, illustrated in Figure 17.3. This means that as

Table 17.2

1	2	3 (SUB)	4	5	6	7	8	9 (PCM)	10	11	12	13	14 (SUB)	15	16

		Frequency	*Angle (deg)	Number of blocks	Time (μs)
1	MARGIN	$1/2\,f_{ch}$	5.051	11	420.9
2	PLL (SUB)	$1/2\,f_{ch}$	0.918	2	76.5
3	SUB-1		3.673	8	306.1
4	POST AMBLE	$1/2\,f_{ch}$	0.459	1	38.3
5	IBG	$1/6\,f_{ch}$	1.378	3	114.8
6	ATF		2.296	5	191.3
7	IBG	$1/6\,f_{ch}$	1.378	3	114.8
8	PLL (PCM)	$1/2\,f_{ch}$	0.918	2	76.5
9	PCM		58.776	128	4898.0
10	IBG	$1/6\,f_{ch}$	1.378	3	114.8
11	ATF		2.296	5	191.3
12	IBG	$1/6\,f_{ch}$	1.378	3	114.8
13	PLL (SUB)	$1/2\,f_{ch}$	0.918	2	76.5
14	SUB-2		3.673	8	306.1
15	POST AMBLE	$1/2\,f_{ch}$	0.459	1	38.3
16	MARGIN	$1/2\,f_{ch}$	5.051	11	420.9
	Total		90	196	7500

Recording density 61.0 Kbit in^{-1}
f_{ch} 9.408 MHz
*Values for 30 mm diameter, 90° wrap angle, 2000 rpm cylinder

much tape as possible is used – rotating head recorders without this overwrite record facility must leave a guardband between each track on the tape. Because of this, recorders using overwrite recording techniques are sometimes known as **guard-bandless**. To prevent crosstalk on playback (as each head is wide enough to pick up all of its own track and half of the next), the heads are set at azimuth angles of ±20°. This enables, as will be explained later, automatic track following (ATF).

These overwrite record and head azimuth techniques are fairly standard approaches to rotating head video recording, and are used specifically to increase the recording density.

Figure 17.4 shows the R-DAT track format on the tape, while Table 17.2 shows the track contents. Table 17.2 lists each part of a track and gives the recording angle, recording period and number of blocks allocated to each part. Frequencies of these blocks which are not of a digital-data form are also listed.

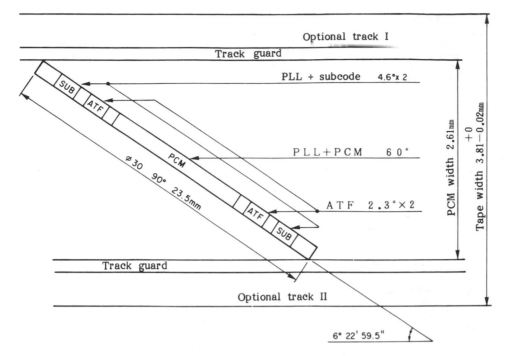

Figure 17.4 *R-DAT tape track format*

As specified in the standard, a head drum with 30 mm diameter is applied and rotates at a 2000 rpm speed. However, in future applications smaller drums with appropriate speeds can be used. At this size and speed, the drum has a resistance to external disturbances similar to that of a gyroscope.

Under these conditions, the 2.46 Mbit s^{-1} signal to be recorded, which includes audio as well as many other types of data, is compressed by a factor of 3 and processed at 7.5 Mbit s^{-1}. This enables the signal to be recorded continuously.

In order to overcome the well-known low frequency problems of coupling transformers in the record/playback head, an 8/10 modulation channel code converts the 8-bit signals to 10-bit signals.

This channel coding also gives the benefit of reducing the range of wave lengths to be recorded. The resultant maximum wave length is only four times the minimum wave length. This allows overwriting, eliminating the need for a separate erase head.

The track outline is given in Figure 17.4. Each helical track is divided into seven areas, separated by interblock gaps. As can be seen, each track has one PCM area, containing the modulated digital information (audio data and error codes), and is 128 blocks of 288 bits long. Table 17.2 lists all track parts of a track.

The PCM area is separated from the other areas by an IBG (inter-block

Table 17.3 *PCM area format*

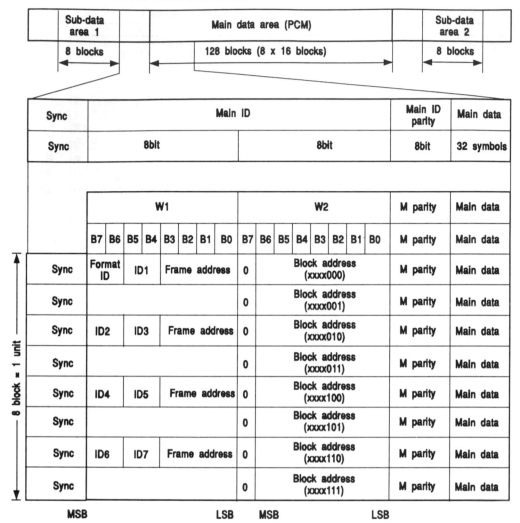

Sync	Main ID		Main ID parity	Main data
Sync	8bit	8bit	8bit	32 symbols

	W1								W2								M parity	Main data
	B7	B6	B5	B4	B3	B2	B1	B0	B7	B6	B5	B4	B3	B2	B1	B0	M parity	Main data
Sync	Format ID		ID1		Frame address			0	Block address (xxxx000)								M parity	Main data
Sync								0	Block address (xxxx001)								M parity	Main data
Sync	ID2		ID3		Frame address			0	Block address (xxxx010)								M parity	Main data
Sync								0	Block address (xxxx011)								M parity	Main data
Sync	ID4		ID5		Frame address			0	Block address (xxxx100)								M parity	Main data
Sync								0	Block address (xxxx101)								M parity	Main data
Sync	ID6		ID7		Frame address			0	Block address (xxxx110)								M parity	Main data
Sync								0	Block address (xxxx111)								M parity	Main data

MSB   LSB MSB   LSB

gap), 3 blocks long. At both sides of the PCM area, two ATF areas are inserted, each 5 blocks long.

Again, an IBG block is inserted at both ends of the track separating the ATF areas from the sub-1 and sub-2 areas (subcode areas), each 8 blocks long. These subareas contain all the information on time code, tape contents, etc.

Then at both track ends a margin block is inserted, 11 blocks long, and is used to cover tolerances in the tape mechanism and head position.

A single track comprises 196 blocks of data, of which the major part is made up of 128 blocks of PCM data. Other important parts are the subcode blocks (sub-1 and sub-2, containing system data, similar to the CD subcode data), automatic track-finding (ATF) signals (to allow high-speed search),

Table 17.4 *Bit assignment of ID-codes*

	Usage	Bit assignment		
ID1	Emphasis	B5	B4	
		0	0:	Off
		0	1:	50/15 μsec
ID2	Sampling frequency	B7	B6	
		0	0:	48 kHz
		0	1:	44.1 kHz
		1	0:	32 kHz
ID3	Number of channel	B5	B4	
		0	0:	2 channels
		0	1:	4 channels
ID4	Quantization	B7	B6	
		0	0:	16 bits linear
		0	1:	12 bits non linear
ID5	Track pitch	B4	B4	
		0	0:	Normal track mode
		0	1:	Wide track mode
ID6	Digital copy	B7	B6	
		0	0:	Permitted
		1	0:	Prohibited
		1	1:	permitted only for the first generation
ID7	Pack	B5	B4:	Pack contents

and the inter-block gaps around the ATF signals (which means that the PCM and subcode information can be overwritten independently without inter-ference to surrounding areas). Parts are recorded successively along the track.

The PCM area format is shown in Table 17.3. PCM and subcode parts comprise similar data blocks, shown in Figure 17.5. Each block is 288 bits long.

Each block comprises 8 synchronization bits, the identification word (W1, 8 bit), the block address word (W2, 8 bit), 8-bit parity word and 256 bits (32 × 8-bit symbol) data. The ID-code W1 contains control signals related to the main data. Table 17.4 shows the bit assignment of the ID codes. The W2 contains the block address. The MSB (most significant bit) of the W2 word

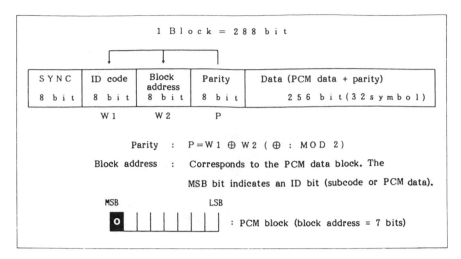

Figure 17.5 *PCM and subcode data blocks*

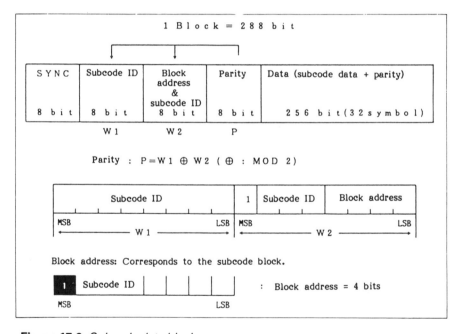

Figure 17.6 *Subcode data blocks*

defines whether the data block is of PCM or subcode form. Where the MSB is zero, the block consists of PCM audio data, and the remainder of word W2, i.e., seven bits, gives the block address within the track. The 7 bits therefore identify the absolute block address (as 2^7 is 128).

On the other hand, when the MSB of word W2 is 1, the block is of subcode form and data bits in the word are as shown in Figure 17.6, where a further 3 bits are used to extend the W1 word subcode identity code, and the four least significant bits give the block address.

The P-word, block parity, is used to check the validity of the W1 and W2 words and is calculated as follows:

$$P = W1 (+) W2$$

where $(+)$ signifies modulo-2 addition as explained in Appendix 1.

Automatic Track Following

In the R-DAT system, no control track is provided. In order to obtain correct tracking during playback, a unique ATF signal is recorded along with the digital data.

The ATF track pattern is illustrated in Figure 17.7. One data frame is completed in two tracks and one ATF pattern completed in two frames (four tracks). Each frame has an A and a B track. A tracks are recorded by the head with $+20°$ azimuth and B tracks are recorded by the head with $-20°$ azimuth.

The ATF signal pattern is repeated over subsequent groups of four tracks. The frequencies of the ATF-signals are listed in Figure 17.7. The key to the operation lies in the fact that different frames hold different combinations and lengths. Furthermore, the ATF operation is based upon the use of the crosstalk signals, picked up by the wide head, which is 1.5 times the track width, and the azimuth recording. This method is called the **area divided ATF**.

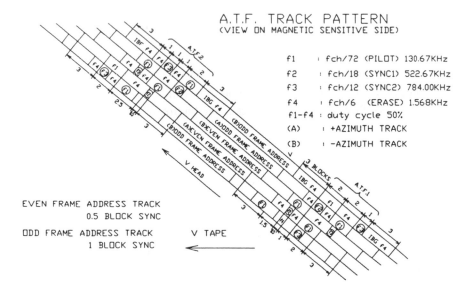

Figure 17.7 *ATF-signal frequencies*

As shown in Figure 17.7, the ATF uses a pilot signal f_1; sync signal 1, f_2; sync signal 2, f_3; and erase signal, f_4. When the head passes along the track in the direction of the arrow (V-head) and detects an f_2 or f_3 signal, the adjacent 6 pilot signals f_1 on both sides are immediately compared, which results in a correction of the tracking when necessary.

The f_2 and f_3 signals thus act as sync signals to start the ATF servo operation.

The f_1 signal, a low frequency signal, i.e., 130.67 kHz, is used as low frequency signals are not affected by the azimuth setting, so crosstalk can be picked up and detected from both sides. The pilot signal f_1 is positioned so not to overlap through the head scans across three successive tracks.

Error Correction

As with any digital recording format, the error-detection and -correction scheme is very important. It must detect and correct the digital audio data, as well as subcodes, ID codes and other auxiliary data.

Types of errors that must be corrected are burst errors: dropouts caused by dust, scratches, and head clogging, and random errors: caused by crosstalk from an adjacent track, traces of an imperfectly erased or overwritten signal, or mechanical instability.

Error-correction strategy

In common with other digital audio systems, R-DAT uses a significant amount of error-correction coding to allow error-free replay of recorded information. The error-correction code used is a double-encoded Reed-Solomon code.

These two Reed-Solomon codes produce C1 (32,28) and C2 (32,26) parity symbols, which are calculated on G_F (2^8) by the polynomial:

$$g(x) = x^8 + x^4 + x^3 + x^2 + 1$$

C1 is interleaved on two neighbouring blocks, while C2 is interleaved on one entire track of PCM data every 4 blocks. See Figure 17.8 for the interleaving format.

In order to perform C1 $\rightleftarrows$ C2 decoding/encoding, one track worth of data must be stored in memory.

One track contains 128 blocks consisting of 4096 (32×128) symbols. Of these, 1184 symbols (512 symbols C1 parity and 672 symbols C2 parity) are used for error correction, leaving 2912 data symbols (24×104).

In fact, C1 encoding adds 4 symbols of parity to the 28 data symbols: C1 (32,28); while C2 encoding adds 6 symbols of parity to every 26 PCM data symbols: C2 (32, 26).

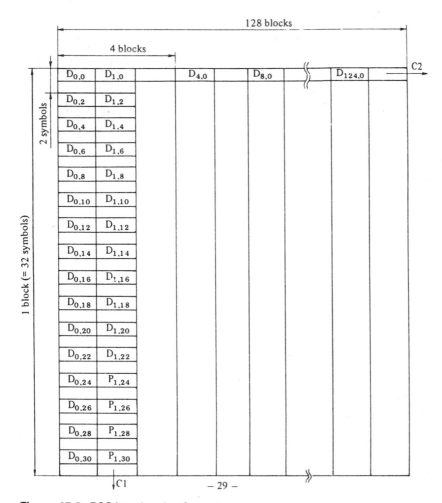

Figure 17.8 *ECC interleaving format*

The main data allocation is shown in Figure 17.9.

This double-Reed-Solomon code gives the format a powerful correction capability for random errors.

PCM data interleave

In order to cope with burst errors, i.e., head clogging, tape dropouts, etc., PCM data is interleaved over two tracks called one frame, effectively turning burst errors into random errors which are correctable using the Reed-Solomon technique already described.

To interleave the PCM data, the contents of two tracks has first to be processed in a memory. The memory size required for one PCM interleave

	0	1	2	3		51	52	53		75	76	77		126	127
0	$D_{0,0}$	$D_{1,0}$	$D_{2,0}$	$D_{3,0}$		$D_{51,0}$	$Q_{52,0}$	$Q_{53,0}$		$Q_{75,0}$	$D_{76,0}$	$D_{77,0}$		$D_{126,0}$	$D_{127,0}$
1	$D_{0,1}$	$D_{1,1}$	$D_{2,1}$	$D_{3,1}$		$D_{51,1}$	$Q_{52,1}$	$Q_{53,1}$		$Q_{75,1}$	$D_{76,1}$	$D_{77,1}$		$D_{126,1}$	$D_{127,1}$
2	$D_{0,2}$	$D_{1,2}$	$D_{2,2}$	$D_{3,2}$		$D_{51,2}$	$Q_{52,2}$	$Q_{53,2}$		$Q_{75,2}$	$D_{76,2}$	$D_{77,2}$		$D_{126,2}$	$D_{127,2}$
3	$D_{0,3}$	$D_{1,3}$	$D_{2,3}$	$D_{3,3}$		$D_{51,3}$	$Q_{52,3}$	$Q_{53,3}$		$Q_{75,3}$	$D_{76,3}$	$D_{77,3}$		$D_{126,3}$	$D_{127,3}$
4	$D_{0,4}$	$D_{1,4}$	$D_{2,4}$	$D_{3,4}$		$D_{51,4}$	$Q_{52,4}$	$Q_{53,4}$		$Q_{75,4}$	$D_{76,4}$	$D_{77,4}$		$D_{126,4}$	$D_{127,4}$
5	$D_{0,5}$	$D_{1,5}$	$D_{2,5}$	$D_{3,5}$		$D_{51,5}$	$Q_{52,5}$	$Q_{53,5}$		$Q_{75,5}$	$D_{76,5}$	$D_{77,5}$		$D_{126,5}$	$D_{127,5}$
6	$D_{0,6}$	$D_{1,6}$	$D_{2,6}$	$D_{3,6}$		$D_{51,6}$	$Q_{52,6}$	$Q_{53,6}$		$Q_{75,6}$	$D_{76,6}$	$D_{77,6}$		$D_{126,6}$	$D_{127,6}$
7	$D_{0,7}$	$D_{1,7}$	$D_{2,7}$	$D_{3,7}$		$D_{51,7}$	$Q_{52,7}$	$Q_{53,7}$		$Q_{75,7}$	$D_{76,7}$	$D_{77,7}$		$D_{126,7}$	$D_{127,7}$
8	$D_{0,8}$	$D_{1,8}$	$D_{2,8}$	$D_{3,8}$		$D_{51,8}$	$Q_{52,8}$	$Q_{53,8}$		$Q_{75,8}$	$D_{76,8}$	$D_{77,8}$		$D_{126,8}$	$D_{127,8}$
9	$D_{0,9}$	$D_{1,9}$	$D_{2,9}$	$D_{3,9}$		$D_{51,9}$	$Q_{52,9}$	$Q_{53,9}$		$Q_{75,9}$	$D_{76,9}$	$D_{77,9}$		$D_{126,9}$	$D_{127,9}$
10	$D_{0,10}$	$D_{1,10}$	$D_{2,10}$	$D_{3,10}$		$D_{51,10}$	$Q_{52,10}$	$Q_{53,10}$		$Q_{75,10}$	$D_{76,10}$	$D_{77,10}$		$D_{126,10}$	$D_{127,10}$
11	$D_{0,11}$	$D_{1,11}$	$D_{2,11}$	$D_{3,11}$		$D_{51,11}$	$Q_{52,11}$	$Q_{53,11}$		$Q_{75,11}$	$D_{76,11}$	$D_{77,11}$		$D_{126,11}$	$D_{127,11}$
12	$D_{0,12}$	$D_{1,12}$	$D_{2,12}$	$D_{3,12}$		$D_{51,12}$	$Q_{52,12}$	$Q_{53,12}$		$Q_{75,12}$	$D_{76,12}$	$D_{77,12}$		$D_{126,12}$	$D_{127,12}$
13	$D_{0,13}$	$D_{1,13}$	$D_{2,13}$	$D_{3,13}$		$D_{51,13}$	$Q_{52,13}$	$Q_{53,13}$		$Q_{75,13}$	$D_{76,13}$	$D_{77,13}$		$D_{126,13}$	$D_{127,13}$
14	$D_{0,14}$	$D_{1,14}$	$D_{2,14}$	$D_{3,14}$		$D_{51,14}$	$Q_{52,14}$	$Q_{53,14}$		$Q_{75,14}$	$D_{76,14}$	$D_{77,14}$		$D_{126,14}$	$D_{127,14}$
15	$D_{0,15}$	$D_{1,15}$	$D_{2,15}$	$D_{3,15}$		$D_{51,15}$	$Q_{52,15}$	$Q_{53,15}$		$Q_{75,15}$	$D_{76,15}$	$D_{77,15}$		$D_{126,15}$	$D_{127,15}$
16	$D_{0,16}$	$D_{1,16}$	$D_{2,16}$	$D_{3,16}$		$D_{51,16}$	$Q_{52,16}$	$Q_{53,16}$		$Q_{75,16}$	$D_{76,16}$	$D_{77,16}$		$D_{126,16}$	$D_{127,16}$
17	$D_{0,17}$	$D_{1,17}$	$D_{2,17}$	$D_{3,17}$		$D_{51,17}$	$Q_{52,17}$	$Q_{53,17}$		$Q_{75,17}$	$D_{76,17}$	$D_{77,17}$		$D_{126,17}$	$D_{127,17}$
18	$D_{0,18}$	$D_{1,18}$	$D_{2,18}$	$D_{3,18}$		$D_{51,18}$	$Q_{52,18}$	$Q_{53,18}$		$Q_{75,18}$	$D_{76,18}$	$D_{77,18}$		$D_{126,18}$	$D_{127,18}$
19	$D_{0,19}$	$D_{1,19}$	$D_{2,19}$	$D_{3,19}$		$D_{51,19}$	$Q_{52,19}$	$Q_{53,19}$		$Q_{75,19}$	$D_{76,19}$	$D_{77,19}$		$D_{126,19}$	$D_{127,19}$
20	$D_{0,20}$	$D_{1,20}$	$D_{2,20}$	$D_{3,20}$		$D_{51,20}$	$Q_{52,20}$	$Q_{53,20}$		$Q_{75,20}$	$D_{76,20}$	$D_{77,20}$		$D_{126,20}$	$D_{127,20}$
21	$D_{0,21}$	$D_{1,21}$	$D_{2,21}$	$D_{3,21}$		$D_{51,21}$	$Q_{52,21}$	$Q_{53,21}$		$Q_{75,21}$	$D_{76,21}$	$D_{77,21}$		$D_{126,21}$	$D_{127,21}$
22	$D_{0,22}$	$D_{1,22}$	$D_{2,22}$	$D_{3,22}$		$D_{51,22}$	$Q_{52,22}$	$Q_{53,22}$		$Q_{75,22}$	$D_{76,22}$	$D_{77,22}$		$D_{126,22}$	$D_{127,22}$
23	$D_{0,23}$	$D_{1,23}$	$D_{2,23}$	$D_{3,23}$		$D_{51,23}$	$Q_{52,23}$	$Q_{53,23}$		$Q_{75,23}$	$D_{76,23}$	$D_{77,23}$		$D_{126,23}$	$D_{127,23}$
24	$D_{0,24}$	$P_{1,24}$	$D_{2,24}$	$P_{3,24}$		$P_{51,24}$	$Q_{52,24}$	$P_{53,24}$		$P_{75,24}$	$D_{76,24}$	$P_{77,24}$		$D_{126,24}$	$P_{127,24}$
25	$D_{0,25}$	$P_{1,25}$	$D_{2,25}$	$P_{3,25}$		$P_{51,25}$	$Q_{52,25}$	$P_{53,25}$		$P_{75,25}$	$D_{76,25}$	$P_{77,25}$		$D_{126,25}$	$P_{127,25}$
26	$D_{0,26}$	$P_{1,26}$	$D_{2,26}$	$P_{3,26}$		$P_{51,26}$	$Q_{52,26}$	$P_{53,26}$		$P_{75,26}$	$D_{76,26}$	$P_{77,26}$		$D_{126,26}$	$P_{127,26}$
27	$D_{0,27}$	$P_{1,27}$	$D_{2,27}$	$P_{3,27}$		$P_{51,27}$	$Q_{52,27}$	$P_{53,27}$		$P_{75,27}$	$D_{76,27}$	$P_{77,27}$		$D_{126,27}$	$P_{127,27}$
28	$D_{0,28}$	$P_{1,28}$	$D_{2,28}$	$P_{3,28}$		$P_{51,28}$	$Q_{52,28}$	$P_{53,28}$		$P_{75,28}$	$D_{76,28}$	$P_{77,28}$		$D_{126,28}$	$P_{127,28}$
29	$D_{0,29}$	$P_{1,29}$	$D_{2,29}$	$P_{3,29}$		$P_{51,29}$	$Q_{52,29}$	$P_{53,29}$		$P_{75,29}$	$D_{76,29}$	$P_{77,29}$		$D_{126,29}$	$P_{127,29}$
30	$D_{0,30}$	$P_{1,30}$	$D_{2,30}$	$P_{3,30}$		$P_{51,30}$	$Q_{52,30}$	$P_{53,30}$		$P_{75,30}$	$D_{76,31}$	$P_{77,30}$		$D_{126,30}$	$P_{127,30}$
31	$D_{0,31}$	$P_{1,31}$	$D_{2,31}$	$P_{3,31}$		$P_{51,31}$	$Q_{52,31}$	$P_{53,31}$		$P_{75,31}$	$D_{76,31}$	$P_{77,31}$		$D_{126,31}$	$P_{127,31}$

1 block (= 32 symbols)

128 blocks

Figure 17.9 *Data allocation*

block is: (128×32) symbols $\times$ 8 bit $\times$ 2 tracks = 65.536 bit, which means a 128 bit memory is required.

The symbols are interleaved, based on the following method, according to the respective number of the audio data symbol. The interleaving format depends on whether a 16-bit or 12-bit quantization is used. The interleave format discussed here is for 16-bit quantization; the most important format.

One 16-bit audio data word indicated as A_i or B_i is converted to two audio data symbols each consisting of 8 bits. The audio data symbol converted from the upper 8 bits of A_i or B_i is expressed as A_{iu} or B_{iu}. The audio data symbol converted from the lower 8 bits of A_i or B_i is expressed as A_{il} or B_{il}.

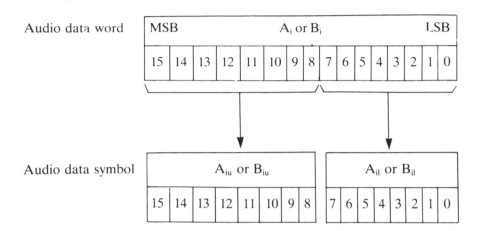

Note: A stands for left channel, B for right channel.

If the audio data symbol is equal to A_{iu} or A_{il}, let a = 0.
If the audio data symbol is equal to B_{iu} or B_{il}, let a = 1.
If the audio data symbol is equal to A_{iu} or B_{iu}, let u = 0.
If the audio data symbol is equal to A_{il} or B_{il}, let u = 1.

Table 17.5 a and b represents an example of the data assignment for both tracks (+ azimuth and − azimuth) respectively, for 16-bit sampled data words.

Subcode

The data subcode capacity is about four times that of a CD and various applications will be available in the future. A subcode format which is essentially the same as the CD subcode format is currently specified for pre-recorded tapes.

The most important control bits, such as the sampling frequency bit and copy inhibit bit, are recorded in the PCM-ID area, so it is impossible to change these bits without rewriting the PCM data. As the PCM data is protected by the main error-correction process, subcodes requiring a high reliability are usefully stored here.

Data to allow fast accessing, programme number, time code, etc., are recorded in subcode areas (sub-1 and sub-2) which are located at both ends of

Table 17.5a

Block address

Recording direction →

Symbol number	0	1	2	3	35	50	51	52	75	76	77	78	79	111	126	127
0	A 0u	A 832u	A 2u	A 834u	A 866u	A 50u	A 882u	Q 52.0	Q 75.0	B 1u	B 833u	B 3u	B 835u	B 867u	B 51u	B 883u
1	A 52u	A 884u	A 54u	A 886u	A 918u	A 102u	A 934u	Q 52.1	Q 75.1	B 53u	B 885u	B 55u	B 887u	B 919u	B 103u	B 935u
2	A 0l	A 832l	A 2l	A 834l	A 866l	A 50l	A 882l	Q 52.2	Q 75.2	B 1l	B 833l	B 3l	B 835l	B 867l	B 51l	B 883l
3	A 52l	A 884l	A 54l	A 886l	A 918l	A 102l	A 934l	Q 52.3	Q 75.3	B 53l	B 885l	B 55l	B 887l	B 919l	B 103l	B 935l
4	A 104u	A 936u	A 106u	A 938u	A 970u	A 154u	A 986u	Q 52.4	Q 75.4	B 105u	B 937u	B 107u	B 939u	B 971u	B 155u	B 987u
5	A 156u	A 988u	A 158u	A 990u	A 1022u	A 206u	A 1038u	Q 52.5	Q 75.5	B 157u	B 989u	B 159u	B 991u	B 1023u	B 207u	B 1039u
6	A 104l	A 936l	A 106l	A 938l	A 970l	A 154l	A 986l	Q 52.6	Q 75.6	B 105l	B 937l	B 107l	B 939l	B 971l	B 155l	B 987l
7	A 156l	A 988l	A 158l	A 990l	A 1022l	A 206l	A 1038l	Q 52.7	Q 75.7	B 157l	B 989l	B 159l	B 991l	B 1023l	B 207l	B 1039l
8	A 208u	A 1040u	A 210u	A 1042u	A 1074u	A 258u	A 1090u	Q 52.8	Q 75.8	B 209u	B 1041u	B 211u	B 1043u	B 1075u	B 259u	B 1091u
9	A 260u	A 1092u	A 262u	A 1094u	A 1126u	A 310u	A 1142u	Q 52.9	Q 75.9	B 261u	B 1093u	B 263u	B 1095u	B 1127u	B 311u	B 1143u
10	A 208l	A 1040l	A 210l	A 1042l	A 1074l	A 258l	A 1090l	Q 52.10	Q 75.10	B 209l	B 1041l	B 211l	B 1043l	B 1075l	B 259l	B 1091l
11	A 260l	A 1092l	A 262l	A 1094l	A 1126l	A 310l	A 1142l	Q 52.11	Q 75.11	B 261l	B 1093l	B 263l	B 1095l	B 1127l	B 311l	B 1143l
12	A 312u	A 1144u	A 314u	A 1146u	A 1178u	A 362u	A 1194u	Q 52.12	Q 75.12	B 313u	B 1145u	B 315u	B 1147u	B 1179u	B 363u	B 1195u
13	A 364u	A 1196u	A 366u	A 1198u	A 1230u	A 414u	A 1246u	Q 52.13	Q 75.13	B 365u	B 1197u	B 367u	B 1199u	B 1231u	B 415u	B 1247u
14	A 312l	A 1144l	A 314l	A 1146l	A 1178l	A 362l	A 1194l	Q 52.14	Q 75.14	B 313l	B 1145l	B 315l	B 1147l	B 1179l	B 363l	B 1195l
15	A 364l	A 1196l	A 366l	A 1198l	A 1230l	A 414l	A 1246l	Q 52.15	Q 75.15	B 365l	B 1197l	B 367l	B 1199l	B 1231l	B 415l	B 1247l
16	A 416u	A 1248u	A 418u	A 1250u	A 1282u	A 466u	A 1298u	Q 52.16	Q 75.16	B 417u	B 1249u	B 419u	B 1251u	B 1283u	B 467u	B 1299u
17	A 468u	A 1300u	A 470u	A 1302u	A 1334u	A 518u	A 1350u	Q 52.17	Q 75.17	B 469u	B 1301u	B 471u	B 1303u	B 1335u	B 519u	B 1351u
18	A 416l	A 1248l	A 418l	A 1250l	A 1282l	A 466l	A 1298l	Q 52.18	Q 75.18	B 417l	B 1249l	B 419l	B 1251l	B 1283l	B 467l	B 1299l
19	A 468l	A 1300l	A 470l	A 1302l	A 1334l	A 518l	A 1350l	Q 52.19	Q 75.19	B 469l	B 1301l	B 471l	B 1303l	B 1335l	B 519l	B 1351l
20	A 520u	A 1352u	A 522u	A 1354u	A 1386u	A 570u	A 1402u	Q 52.20	Q 75.20	B 521u	B 1353u	B 523u	B 1355u	B 1387u	B 571u	B 1403u
21	A 572u	A 1404u	A 574u	A 1406u	A 1438u	A 622u		Q 52.21	Q 75.21	B 573u	B 1405u	B 575u	B 1407u	B 1439u	B 623u	
22	A 520l	A 1352l	A 522l	A 1354l	A 1386l	A 570l	A 1402l	Q 52.22	Q 75.22	B 521l	B 1353l	B 523l	B 1355l	B 1387l	B 571l	B 1403l
23	A 572l	A 1404l	A 574l	A 1406l	A 1438l	A 622l		Q 52.23	Q 75.23	B 573l	B 1405l	B 575l	B 1407l	B 1439l	B 623l	
24	A 624u	P 1.24	A 626u	P 3.24	P 35.24	A 674u	P 51.24	Q 52.24	P 75.24	B 625u	P 77.24	B 627u	P 79.24	P 111.24	B 675u	P 127.24
25	A 676u	P 1.25	A 678u	P 3.25	P 35.25	A 726u	P 51.25	Q 52.25	P 75.25	B 677u	P 77.25	B 679u	P 79.25	P 111.25	B 727u	P 127.25
26	A 624l	P 1.26	A 626l	P 3.26	P 35.26	A 674l	P 51.26	Q 52.26	P 75.26	B 625l	P 77.26	B 627l	P 79.26	P 111.26	B 675l	P 127.26
27	A 676l	P 1.27	A 678l	P 3.27	P 35.27	A 726l	P 51.27	Q 52.27	P 75.27	B 677l	P 77.27	B 679l	P 79.27	P 111.27	B 727l	P 127.27
28	A 728u	P 1.28	A 730u	P 3.28	P 35.28	A 778u	P 51.28	Q 52.28	P 75.28	B 729u	P 77.28	B 731u	P 79.28	P 111.28	B 779u	P 127.28
29	A 780u	P 1.29	A 782u	P 3.29	P 35.29	A 830u	P 51.29	Q 52.29	P 75.29	B 781u	P 77.29	B 783u	P 79.29	P 111.29	B 831u	P 127.29
30	A 728l	P 1.30	A 730l	P 3.30	P 35.30	A 778l	P 51.30	Q 52.30	P 75.30	B 729l	P 77.30	B 731l	P 79.30	P 111.30	B 779l	P 127.30
31	A 780l	P 1.31	A 782l	P 3.31	P 35.31	A 830l	P 51.31	Q 52.31	P 75.31	B 781l	P 77.31	B 783l	P 79.31	P 111.31	B 831l	P 127.31

Recording direction

Symbol number

Table 17.5b

Block address

Symbol number	0	1	2	3	35	50	51	52	75	76	77	78	79	111	126	127
0	B 0u	B 832u	B 2u	B 834u	B 866u	B 50u	B 882u	Q 52.0	Q 75.0	A 1u	A 833u	A 3u	A 835u	A 867u	A 51u	A 883u
1	B 52u	B 884u	B 54u	B 886u	B 918u	B 102u	B 934u	Q 52.1	Q 75.1	A 53u	A 885u	A 55u	A 887u	A 919u	A 103u	A 935u
2	B 01	B 8321	B 21	B 8341	B 8661	B 501	B 8821	Q 52.2	Q 75.2	A 11	A 8331	A 31	A 8351	A 8671	A 511	A 8831
3	B 521	B 8841	B 541	B 8861	B 9181	B 1021	B 9341	Q 52.3	Q 75.3	A 531	A 8851	A 551	A 8871	A 9191	A 1031	A 9351
4	B 104u	B 936u	B 106u	B 938u	B 970u	B 154u	B 986u	Q 52.4	Q 75.4	A 105u	A 937u	A 107u	A 939u	A 971u	A 155u	A 987u
5	B 156u	B 988u	B 158u	B 990u	B 1022u	B 206u	B 1038u	Q 52.5	Q 75.5	A 157u	A 989u	A 159u	A 991u	A 1023u	A 207u	A 1039u
6	B 1041	B 9361	B 1061	B 9381	B 9701	B 1541	B 9861	Q 52.6	Q 75.6	A 1051	A 9371	A 1071	A 9391	A 9711	A 1551	A 9871
7	B 1561	B 9881	B 1581	B 9901	B 10221	B 2061	B 10381	Q 52.7	Q 75.7	A 1571	A 9891	A 1591	A 9911	A 10231	A 2071	A 10391
8	B 208u	B 1040u	B 210u	B 1042u	B 1074u	B 258u	B 1090u	Q 52.8	Q 75.8	A 209u	A 1041u	A 211u	A 1043u	A 1075u	A 259u	A 1091u
9	B 260u	B 1092u	B 262u	B 1094u	B 1126u	B 310u	B 1142u	Q 52.9	Q 75.9	A 261u	A 1093u	A 263u	A 1095u	A 1127u	A 311u	A 1143u
10	B 2081	B 10401	B 2101	B 10421	B 10741	B 2581	B 10901	Q 52.10	Q 75.10	A 2091	A 10411	A 2111	A 10431	A 10751	A 2591	A 10911
11	B 2601	B 10921	B 2621	B 10941	B 11261	B 3101	B 11421	Q 52.11	Q 75.11	A 2611	A 10931	A 2631	A 10951	A 11271	A 3111	A 11431
12	B 312u	B 1144u	B 314u	B 1146u	B 1178u	B 362u	B 1194u	Q 52.12	Q 75.12	A 313u	A 1145u	A 315u	A 1147u	A 1179u	A 363u	A 1195u
13	B 364u	B 1196u	B 366u	B 1198u	B 1230u	B 414u	B 1246u	Q 52.13	Q 75.13	A 365u	A 1197u	A 367u	A 1199u	A 1231u	A 415u	A 1247u
14	B 3121	B 11441	B 3141	B 11461	B 11781	B 3621	B 11941	Q 52.14	Q 75.14	A 3131	A 11451	A 3151	A 11471	A 11791	A 3631	A 11951
15	B 3641	B 11961	B 3661	B 11981	B 12301	B 4141	B 12461	Q 52.15	Q 75.15	A 3651	A 11971	A 3671	A 11991	A 12311	A 4151	A 12471
16	B 416u	B 1248u	B 418u	B 1250u	B 1282u	B 466u	B 1298u	Q 52.16	Q 75.16	A 417u	A 1249u	A 419u	A 1251u	A 1283u	A 467u	A 1299u
17	B 468u	B 1300u	B 470u	B 1302u	B 1334u	B 518u	B 1350u	Q 52.17	Q 75.17	A 469u	A 1301u	A 471u	A 1303u	A 1335u	A 519u	A 1351u
18	B 4161	B 12481	B 4181	B 12501	B 12821	B 4661	B 12981	Q 52.18	Q 75.18	A 4171	A 12491	A 4191	A 12511	A 12831	A 4671	A 12991
19	B 4681	B 13001	B 4701	B 13021	B 13341	B 5181	B 13501	Q 52.19	Q 75.19	A 4691	A 13011	A 4711	A 13031	A 13351	A 5191	A 13511
20	B 520u	B 1352u	B 522u	B 1354u	B 1386u	B 570u	B 1402u	Q 52.20	Q 75.20	A 521u	A 1353u	A 523u	A 1355u	A 1387u	A 571u	A 1403u
21	B 572u	B 1404u	B 574u	B 1406u	B 1438u	B 622u	B 1454u	Q 52.21	Q 75.21	A 573u	A 1405u	A 575u	A 1407u	A 1439u	A 623u	A 1455u
22	B 5201	B 13521	B 5221	B 13541	B 13861	B 5701	B 14021	Q 52.22	Q 75.22	A 5211	A 13531	A 5231	A 13551	A 13871	A 5711	A 14031
23	B 5721	B 14041	B 5741	B 14061	B 14381	B 6221	B 14541	Q 52.23	Q 75.23	A 5731	A 14051	A 5751	A 14071	A 14391	A 6231	A 14551
24	B 624u	P 1.24	B 626u	P 3.24	P 35.24	B 674u	P 51.24	Q 52.24	P 75.24	A 625u	P 77.24	A 627u	P 79.24	P 111.24	A 675u	P 127.24
25	B 676u	P 1.25	B 678u	P 3.25	P 35.25	B 726u	P 51.25	Q 52.25	P 75.25	A 677u	P 77.25	A 679u	P 79.25	P 111.25	A 727u	P 127.25
26	B 6241	P 1.26	B 6261	P 3.26	P 35.26	B 6741	P 51.26	Q 52.26	P 75.26	A 6251	P 77.26	A 6271	P 79.26	P 111.26	A 6751	P 127.26
27	B 6761	P 1.27	B 6781	P 3.27	P 35.27	B 7261	P 51.27	Q 52.27	P 75.27	A 6771	P 77.27	A 6791	P 79.27	P 111.27	A 7271	P 127.27
28	B 728u	P 1.28	B 730u	P 3.28	P 35.28	B 778u	P 51.28	Q 52.28	P 75.28	A 729u	P 77.28	A 731u	P 79.28	P 111.28	A 779u	P 127.28
29	B 780u	P 1.29	B 782u	P 3.29	P 35.29	B 830u	P 51.29	Q 52.29	P 75.29	A 781u	P 77.29	A 783u	P 79.29	P 111.29	A 831u	P 127.29
30	B 7281	P 1.30	B 7301	P 3.30	P 35.30	B 7781	P 51.30	Q 52.30	P 75.30	A 7291	P 77.30	A 7311	P 79.30	P 111.30	A 7791	P 127.30
31	B 7801	P 1.31	B 7821	P 3.31	P 35.31	B 8301	P 51.31	Q 52.31	P 75.31	A 7811	P 77.31	A 7831	P 79.31	P 111.31	A 8311	P 127.31

Recording direction →

Symbol number

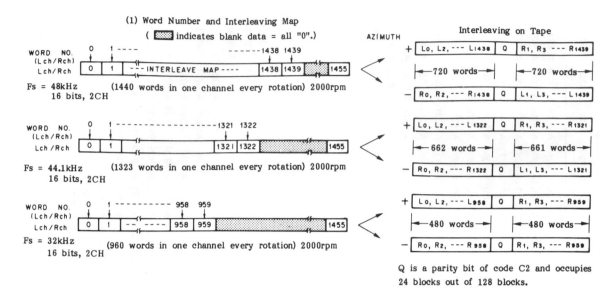

Q is a parity bit of code C2 and occupies
24 blocks out of 128 blocks.

Figure 17.10 *PCM data interleave format*

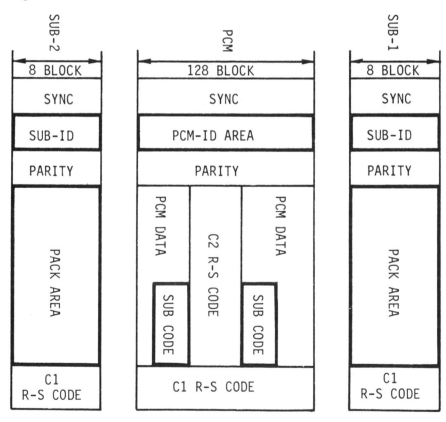

Figure 17.11

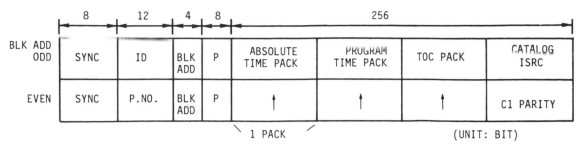

Figure 17.12 *Subcode area format*

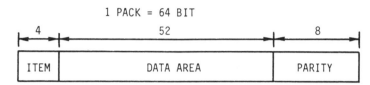

Figure 17.13 *PACK format*

Table 17.6

Item	Mode
0000	No information
0001	Programme time
0010	Absolute time
0011	Running time
0100	TOC
0101	Calendar
0110	Catalogue
0111	ISRC
1000	Reserved
1110	
1111	For tape maker

the helical tracks. These subcode areas are identical. Figure 17.11 illustrates the sub-1 and sub-2 areas, along with the PCM area containing subcode information.

An example of the subcode area format is shown in Figure 17.12. Data are recorded in a **pack** format.

Figure 17.13 shows the pack format, and the pack item codes are listed in Table 17.6. All the CD-Q channel subcodes are available to be used.

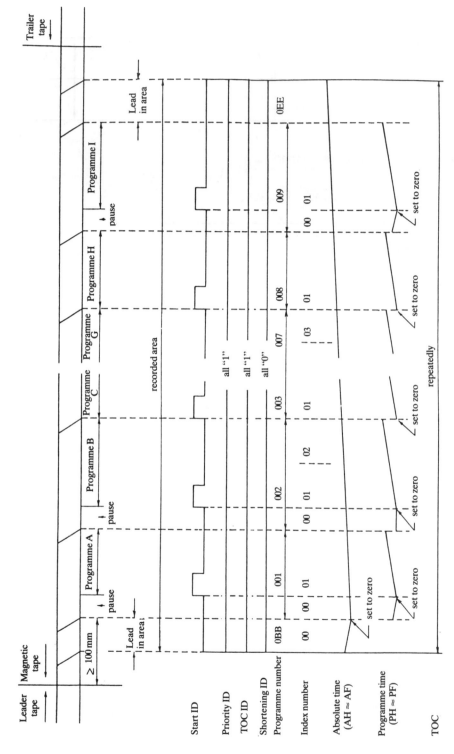

Figure 17.14

Each pack block comprises an item code of 4 bits, indicating what informa-tion is stored in the pack data area. The item code 0100 indicates that the related pack data is a TOC (table of contents) pack. This TOC is recorded repeatedly throughout the tape, in order to allow high-speed access and search (at 200 × normal speed). Every subcode datablock is controlled by an 8-bit C1 parity word allowing appropriate control of data validity.

Subcode data in the subarea can be rewritten or modified independently from the PCM data.

Figure 17.14 shows an example of subcode information for pre-recorded tape. The figure shows the use of different codes and pack data on a tape, such as programme time, absolute time, programme number, etc.

Tape Duplication

High-speed duplication of R-DAT tapes can be done by using the magnetic contact printing technique. In this method a master tape of the mirror type is produced on a master tape recorder (Figure 17.15a).

The magnetic surfaces of the master tape and the copy tape are mounted in contact with each other on a printing machine as shown in Figure 17.15b.

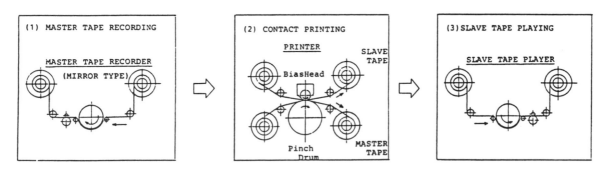

Figure 17.15

By controlling the pressure of both tapes between pinch drum and bias head, the magnetizing process is performed, applying a magnetic bias to the contact area. Special tape and a special bias head are required (see Figures 17.16 and 17.17).

Cassette

The cassette is a completely sealed structure and measures $73 \times 54 \times 10.5$ mm. It weighs about 20 g. See Figure 17.18.

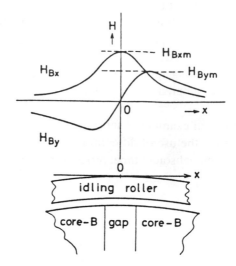

Figure 17.16

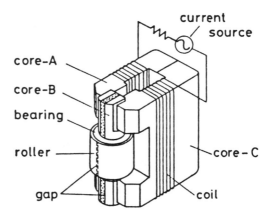

Figure 17.17

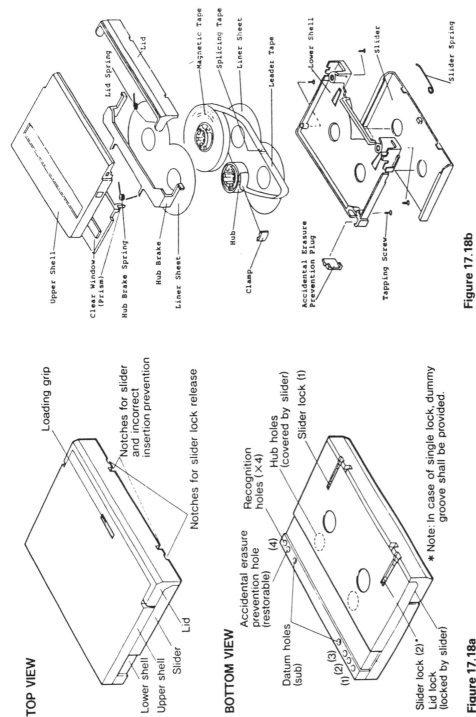

TOP VIEW

Loading grip

Notches for slider and incorrect insertion prevention

Notches for slider lock release

Lower shell
Upper shell
Slider
Lid

BOTTOM VIEW

Accidental erasure prevention hole (restorable)

Recognition holes (×4)

Hub holes (covered by slider)

Slider lock (1)

(4)

Datum holes (sub)

(1) (2) (3)

Slider lock (2)

Lid lock (locked by slider)

* Note: In case of single lock, dummy groove shall be provided.

Figure 17.18a

Lid Spring

Lid

Magnetic Tape

Splicing Tape

Liner Sheet

Leader Tape

Upper Shell

Clear Window (Prism)

Hub Brake Spring

Hub Brake

Liner Sheet

Hub

Clamp

Lower Shell

Slider

Slider Spring

Accidental Erasure Prevention Plug

Tapping Screw

Figure 17.18b

(Lid closed)

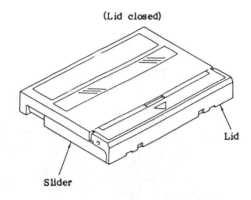

Lid

Slider

(Lid open)

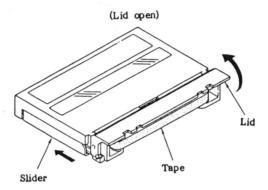

Lid

Tape

Slider

Figure 17.18c

18
Non-tracking Digital Audio Tape (NT-DAT)

A further evolution from the first (tracking) DAT is the NT-DAT.

The advantage of this system is the much improved vibration and shock handling capacity. This technology is used in playback-only units such as car DAT players, but also in a miniature version, the so-called stamp-size DAT player. In the last case, recording with NT-method is also applicable.

NT-Playback

The original DAT specifications are based on the fact that each re/pb head has its own azimuth and that each head must run correctly centred over the correct data track, for which purpose the automatic tracking following (ATF) system is included (see Chapter 17).

Whenever vibration occurs, there is the danger that the head loses track and that recovering the correct position immediately is impossible, resulting in loss of read-out or recorded data. Non-tracking DAT does not use ATF tracking, in fact, there is no tracking at all.

The combination of double density scan, high processor capacity and high memory capacity can overcome momentary loss of tracking.

Double Density Scan

Figure 18.1 shows a comparison of playback waveforms obtained from a 30 mm diameter drum and a 15 mm diameter drum, respectively, however, the drum speed of the latter is 4000 rpm instead of the conventional 2000 rpm. On the other hand, the relative tape to head speed remains the same:

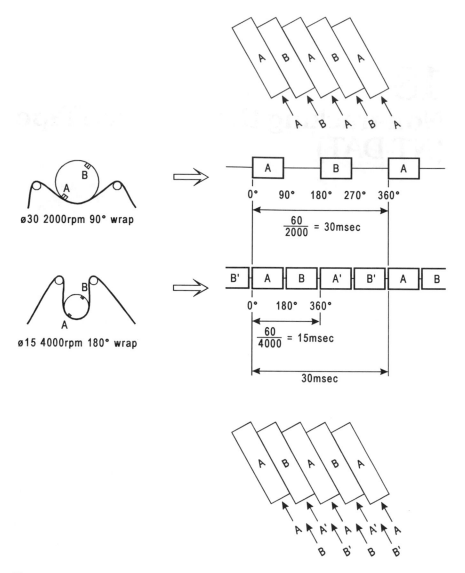

Figure 18.1 *Double density scan*

$$V_r = 30 \times \pi \times \frac{2000}{60} = 15 \times \pi \times \frac{4000}{60} = 3.1 \text{ m/s}$$

In case of the 15 mm drum, the wrap angle is 180°.

Each head will in this case perform read-out, but twice that of the 30 mm drum (the drum speed being doubled), and also not correctly over the centre of the track. The result on the RF level is shown in Fig. 18.2.

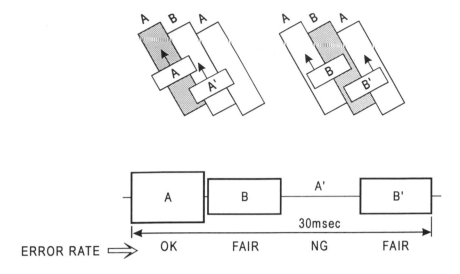

(RF WAVEFORM WHICH HAS BEEN PLAYED BACK)

Figure 18.2 *Head vs. tape*

Depending on the azimuth of the head versus track, it is possible that some read-out is good, fair or not good at all, but in any case, all data will be read out at least one time with sufficient RF level.

If the RF level becomes low, it is obvious that the error rate of the data will increase; this is recovered partly by the error correction and partly by the double read-out.

Another important detail is that the head width in conventional DAT players is 20 μm, whereas it is 28.5 μm in this type of player, more RF is therefore obtained.

In long-play mode, the drum speed will be the same and the tape speed will be halved. This results in a 4-times read-out, hereby decreasing the error rate significantly.

NT-Readout

As the data which is recorded on the tape contains address data (frame address and block address), it is possible to handle this data correctly even if it is retrieved in a non-logical or discontinuous sequence.

As shown in Fig. 18.3, when the tracking angle of the playback head deviates from the recorded track, the read-out order will be non-logical and maybe even discontinuous. This is not a problem because of the frame and block addresses of each data block.

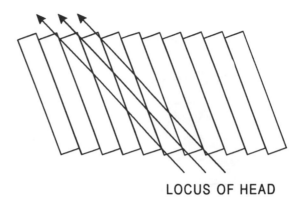

Figure 18.3 *Head locus*

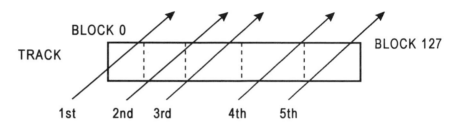

Figure 18.4 *Track readout*

Each data block that has been read out correctly will be stored in random access memory (RAM). After several head read-outs all data blocks of each track will have been read out at least once and stored; if so, the data can be released.

It is obvious that the RAM capacity now becomes important. Supposing an RAM of 1 Mbit is used; this has a capacity of about 12 frames, or 24 tracks, as each frame consists of 2 tracks.

Each track contains 128 blocks of 32 symbols, each symbol contains 8 bits, so each track represents $128 \times 32 \times 8 = 32\,768$ bits (or 32 Kbit), 24 tracks represent 32 Kbit $\times\ 24 = 768$ Kbit.

The difference between 768 Kbit and 1 Mbit is needed for normal processing buffer such as for error processing and others.

NT Stamp-size DAT

NT stamp-size DAT is a miniature DAT format. The dimensions of the cassette are 30 mm $\times$ 5 mm and it weighs 2.8 g.

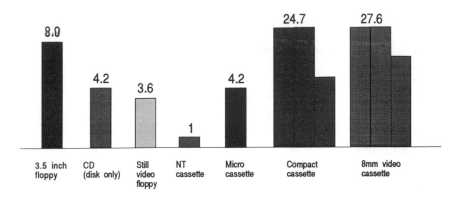

Figure 18.5 *Cassette size comparison*

In this section we will review the essential differences between the conventional DAT and the stamp-size DAT (also referred to as Scoopman).

The size as compared with other types of cassette is extremely small, and yet, it is a digital recordable medium allowing high quality. The sampling rate is 32 kHz, and quantization is 12-bit non-linear, which is comparable to a normal DAT recording in long-play mode.

Table 18.1 *General specifications for NT format*

Recording format	Revolving 2 heads, helical scan, azimuth recording
Tape width	2.5 mm
Tape speed	6.35 mm/s
Drum diameter	14.8 mm
Drum wrap angle	100 degrees (mechanical wrap angle)
Field frequency	50 Hz
Number of drum rotations	50 Hz (double density scan) = 3000 rpm
Relative head speed	Approx. 2.3 mm/s (double density scan)
Track width	Approx. 9.8 μm
Still angle	Approx. 4.4°
Input channels	2 channels stereo
Sampling frequency	32 kHz
Quantization	12-bits non-linear (equivalent to 17 bits)
Error correction	Cross interleave code
Redundancy	33% (including synchronization word, address)
Modulation	Low deviation modulation (LDM)-2

Table 18.2 *Main specifications for NT recorder*

Frequency response	10–14 500 Hz (+1 dB, −3 dB)
Dynamic range	Over 80 dB
S/N	Over 80 dB
Total distortion	Below 0.05 %
Wow and flutter	Below measurement limits (quartz precision)
Dimensions (width × height × depth)	115 mm × 50 mm × 21 mm
Weight	138 g (including battery and cassette)
Power supply	1.5 VDC
Power consumption	Approx. 270 mW (during battery operation)
Lifetime of battery	Approx. 6 h (for recording and playing; using size AA alkaline battery LR6)
Head	Metal in gap (MIG) head
LSI, IC	6 newly developed, 3 widely used

Table 18.3 *General specifications for NT cassette*

Dimension (width × height × depth)	30 mm × 21.5 mm × 5 mm (with lid)
Weight	2.8 g (120-min tape)
Tape:	
Width	2.5 mm
Thickness	Approx. 5 μm
Length	Approx. 20 m (120-min tape)
Type	Ni–Co metal evaporated tape
Maximal recording time	120 min (round trip)
Maximal recording capacity	Approx. 690 mbytes

Double-sided, bi-directional use

Contrary to the conventional DAT type, the stamp-size type is bi-directional; in other words, it is used more like an analog cassette type which has to be reversed.

Tracks are divided into top and bottom halves. In the future, it will be possible to use a bi-directional turnaround mechanism, which, in combination with a large memory size, will enable a smooth, glitchless autoreverse operation.

The tape format is shown in Fig. 18.6; tape width is 2.5 mm, still angle of each track is approximately 4.4° and track pitch is 9.8 µm.

Non-loading method

All the previously used drum-based systems are based on the idea that the tape has to be taken out of its cassette and wrapped around the drum.

In this case however, the technology is reversed. The drum assembly will be inserted into the cassette (of which the front cover is tilted away). Figure 18.7 shows the drum versus the cassette.

This solution allows a design which is much simpler, smaller and more cost-effective, as the loading mechanism is extremely simplified.

One of the major differences is that much of the technology which used to be built into the recorder/playback set is now built into each cassette itself. This has also a very positive effect on the wear of the set. Figure 18.8 shows the non-loading method.

Each cassette contains built-in pinch rollers (also contrary to any conventional design). When inserting the drum into the cassette, the capstan, which

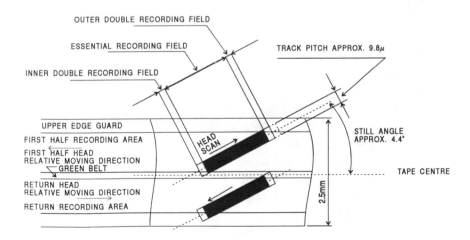

Figure 18.6 *Tape format*

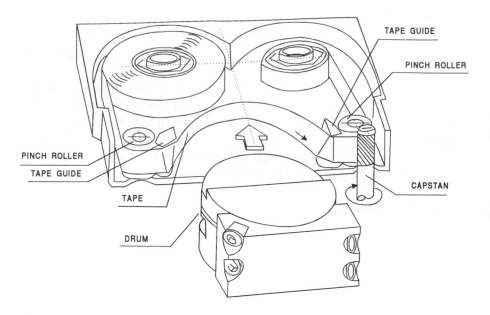

Figure 18.7 *Drum vs. cassette*

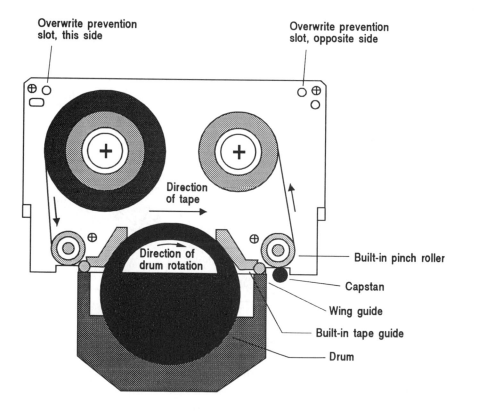

Figure 18.8 *Non-loading method*

is mounted on the same platform as the drum, will automatically be positioned against the pinch roller.

Each cassette also contains tape guides, so that the tape is automatically guided correctly over the inserted drum. Insertion of the drum assembly into the cassette can be performed automatically as well as manually.

Non-tracking Scoopman

In order to attain correct operation, the non-tracking method was chosen, as it was extremely difficult to ensure correct tracking on such a small tape. Note also that a special metal evaporated tape is used to ensure high recording densities, and a high carrier to noise level (c/n).

As already explained, there is no actual tracking; the head will run quite freely over the tape, in this case the drum speed is 3000 rpm. The result is shown in Fig. 18.9, which presents an example of a two-head-drum.

Due to the fact that there is no real tracking, and also due to the double drum speed, the head traces over the tracks in a slanted way.

As each track will be read out several times, though only partially, it is a matter of checking which data was correct, combining it correctly and filling the RAM.

NT-Recording

During recording there is also no tracking; tracks are recorded as they are. The tape speed at that moment is also constant. It is therefore, it is at that moment perfectly possible that because of tape speed fluctuations, vibrations or mechanical tolerances tracks are recorded while overlapping each other slightly, or with a slightly incorrect track angle.

If tracks overlap, it happens at top or bottom, but the data format provides repeat areas at top and bottom, i.e., a fixed number of first and last data blocks are written twice and an eventual overlap can be recovered in this way. Figure 18.10 shows an example of recorded tracks. Drum speed was noted to be 3000 rpm (equals 50 Hz) both in playback and during recording. Due to the specific head use, however, read-out during playback is doubled compared with write during recording. Figure 18.11 shows the drum-head system.

The drum contains two head units, mounted opposite each other, but one of the two head units contains two heads.

In total, we have three heads:

● Head A – recording only, azimuth 27°
● Head A – playback only, azimuth 27°
● Head B – recording/playback, azimuth 27°

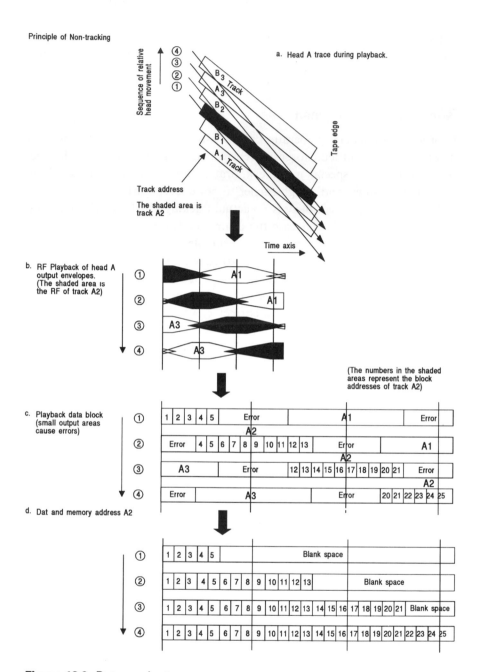

Figure 18.9 *Data readout*

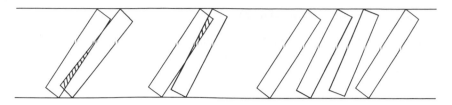

Figure 18.10 *Recorded tracks*

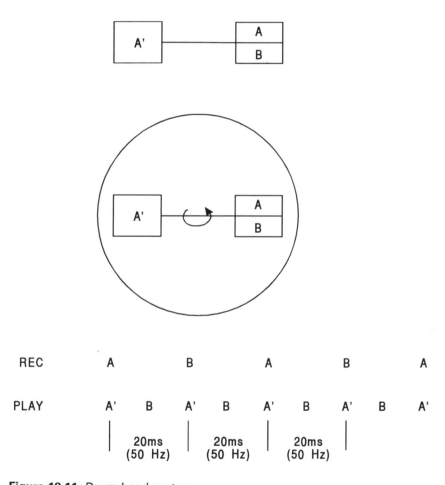

Figure 18.11 *Drum-head system*

Another benefit from this head configuration is the fact that recording is performed by one head unit only. This avoids problems coming from the use of two recording heads, physically different and opposite each other.

Data Format of Stamp-size DAT

Each track contains an essential recording field of about 92 blocks, each block containing 288 bits. Figure 18.12 shows the data format.

Besides this essential field, there are inner and outer double recording fields (where the first and last blocks of the essential field are repeated); at each end there is also a number of unusable blocks which are used for the switchover of heads. The total number of blocks is 116.

In the centre of each track there is a sequence of four control (CTL), one interband gap (IBG), two auxiliaries (AUX), one IBG and four CTL blocks. The use of IBG and AUX is similar to the subcode (SUB) data fields in conventional DAT, this area might be re-recorded without changing the actual music data.

The reason why this part is in the centre of each track is that if it is placed in the centre, it can also be read out even when the tape is in fast forward (FF) and when the tape tension is causing a wrap angle which is smaller than the normal 100° (during FF it is only about 50°).

Within each block, consisting of 288 bits, there are 11 sync bits (always the same pattern), 13 address bits, 16 data words of 12 bits each, four parity words of 12 bits each and 24 cyclic redundancy check (CRC) bits.

The CRC bits are used to detect errors; thus, a block can be declared safe or unsafe, i.e. usable or not usable. The CRC check is performed upon the whole block (253 bits).

The P and Q parity words are even parity checks used to perform error correction; there are P and Q words retrieved from even or odd data samples.

● P parity is even parity based upon 8 data words plus the P word itself
● Q parity is even parity based upon 8 data words plus one P word plus the Q word itself.

It should be noted that P and Q parity works in an interleaved way: each P and Q parity works on data words which are to be found not within the same block, but within the same track. In this way a simple parity check can become a powerful error correction too.

LDM-2 Modulation

In order to attain high-density digital recordings, it is necessary to lengthen the minimum flux reversal interval. It is also necessary to shorten the maximum flux reversal to allow the rotary transformer on the drum to operate without distorsion. Furthermore, the DC component should be as low as possible.

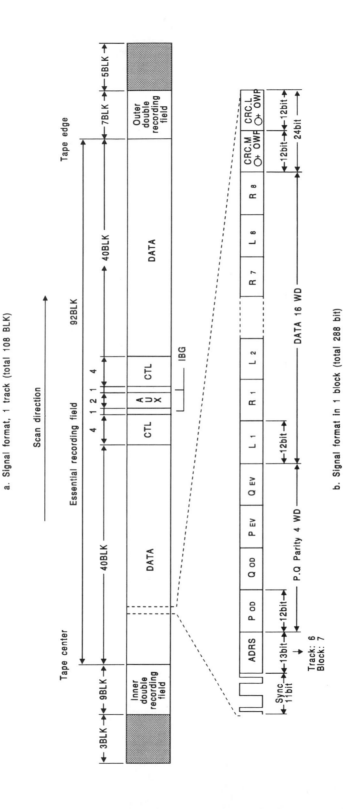

Figure 18.12 *Data format*

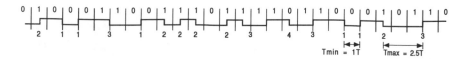

Figure 18.13 *LDM-2 modulation*

The LDM-2 modulation was developed to satisfy these needs and is based on modified frequency modulation (MDM) algorithms as used for floppy disc drives.

The minimum reversal interval is 1 T, the maximum is 2.5 T (1 T is the equivalent of a 1-bit interval before modulation, the basic frequency being 6.144 MHz). Figure 18.13 shows an LDM-2 modulation.

There are four possibilities.

● A transition occurs at the edge of the bit between '0'
● A transition occurs in the middle of the bit when it is '1'
● When an even number of '1' bits occur, the transition between the two last '1' bits occurs at the edge, not in the centre
● If there has been an even number of '1' bits followed by a single '0' bit, the transition in the '0' bit will be in the centre.

19
MiniDisc

MiniDisc (MD) is the latest in the evolution of the disc format. The development target was to produce a disc smaller than conventional discs, easy to use and with recording capacities. Photo 19.1 shows such an MD player.

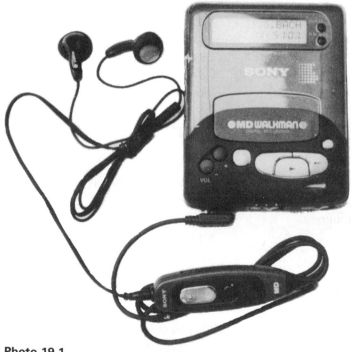

Photo 19.1

The MiniDisc specifications are based upon CD, CD-ROM and CD-MO specifications. Some of the technology is also taken over from these existing formats, other parts were new developments. In this Chapter, the new technology will be explained.

The basic specifications are presented in Table 19.1.

Disc Types

Two disc types are currently available: non-recordable and recordable; both are disc formats which have been mounted into a cartridge for improved safety. Photo 19.2 shows one example of each.

Table 19.1 *MD system specifications*

Main parameters	
Playing and recording time	74 min max.
Cartridge size	68D 72W 5Hmm
Disc parameters	
Diameter	64 mm
Thickness	1.2 mm
Centre hole diameter	11 mm
Starting diameter of lead in area	29 mm max.
Outer diameter of lead in area	32 mm
Diameter of program end	61 mm max.
Track pitch	1.6 µm
Scanning velocity	1.2–1.4 m/s
Audio performance	
Channels	Stereo and mono
Wow and flutter	Quartz crystal precision
Signal format	
Sampling frequency	44.1 kHz
Coding	Adaptive transform coding (ATRAC)
Modulation	Eight-to-fourteen modulation
Error correction system	Advanced CIRC (ACIRC)
Optical parameters	
Laser wavelength	780 nm (typical value)
Lens numerical aperture	0.45 (typical value)
Recording power	2.5–5 mw (main beam)
Recording strategy	Magnetic field modulation

Photo 19.2 *MiniDiscs. Left, non-recordable; right, recordable*

The non-recordable type is very similar to the normal CD type, but the physical format is smaller, yet still contains 74 min of music (ATRAC compressed).

The recordable type is a magneto-optical disc.

Shock-proof capacity

One of the major features of MD format is its shock-proof capacity; the combination of data compression and buffer RAM enables us to have correct music read-out for several seconds when the system is undergoing vibration or heavy shocks causing interruption of read-out. The simple explanation is that data is read out much quicker than needed, it is stored in RAM, as long as there is sufficient data, the player can even go momentarily in pause. Figure 19.3 illustrates the shock-proof capacity of the MD format.

Data read-out speed is the same as that of a CD player, viz 1.4 Mbit/s. However, the decoder only needs 0.3 Mbit/s so there is a buffer RAM which is used to store data temporarily. Each Mbit of RAM allows a buffer capacity of about 3 s. Whenever there is too much input data, i.e. when the RAM is full, the input is stopped by pausing the playback.

On the other hand, if input from the disc should stop momentarily, the buffer RAM will still contain enough data to recover and start again.

Recordability

A recordable disc is not a new development and has been used for years; it is known as CD-magneto-optical (CD-MO). The MO medium for MiniDisc, however, is a further development, allowing a much simpler setup and less recording power.

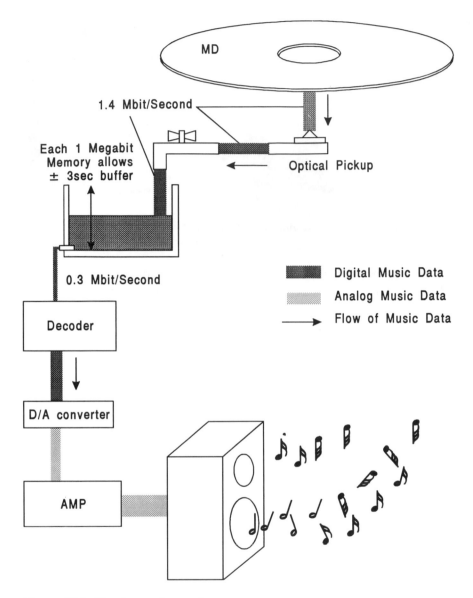

Figure 19.1 *Shock-proof capacity*

Disc recording can be performed with several methods, but the one chosen for MD, the magnetic field modulation (MFM), is the most durable and enables near endless re-recording possibilities (re-recording capacity specified to be 1 000 000 times). Figure 19.2 shows a disc cross-section.

The recordable disc consists of a polycarbonate substrate, on which an MO layer has been laid, on both sides covered by a dielectric layer. On top of this, there is a reflective layer and another protective layer and a silicone lubricant on which the magnetic head will slide.

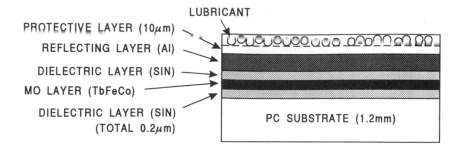

PROTECTIVE LAYER (10μm)
REFLECTING LAYER (Al)
DIELECTRIC LAYER (SIN)
MO LAYER (TbFeCo)
DIELECTRIC LAYER (SIN)
(TOTAL 0.2μm)

LUBRICANT

PC SUBSTRATE (1.2mm)

MEDIA REQUIREMENTS

CONTACT DURABILITY 2×10^6
FRICTION FORCE   LESS THAN 6mN (WEIGHT 20mN)
RECORDABLE TEMPERATURE   - 25°C ~ 70°C

Figure 19.2 *Disc cross-section*

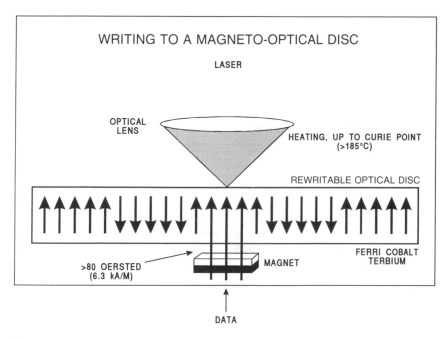

WRITING TO A MAGNETO-OPTICAL DISC

LASER

OPTICAL
LENS

HEATING, UP TO CURIE POINT
(>185°C)

REWRITABLE OPTICAL DISC

FERRI COBALT
TERBIUM

>80 OERSTED
(6.3 kA/M)

MAGNET

DATA

Figure 19.3 *Recording on MD*

The MO layer is a special alloy, ferri-terbium-cobalt (FeTbCo), which has a very low coercivity of about 80 Oersted (6.4kA/m); this means that a low magnetic field is required to store data on such an alloy.

Recording is a combination of a laser and a magnetical head, both on the same axis, each on a side of the disc (opposite each other).

In order to make a magnetic recording on the MO alloy, not only a magnetic field is needed, but also a certain temperature, called the Curie point. In case of FeTbCo, the Curie point is about 185°C. Note that in this way occasional and erroneous erasing of data becomes nearly impossible since recording always needs the combination of the correct temperature and a magnetic field.

The laser will heat up the correct spot and opposite this spot, the magnetic head will put the correct data (+ or −, or North and South, equalling '1' and '0') onto the magnetic layer. A special magnetic head was developed which enables quick flux reversals (approximately 100 ns).

Recording is performed by overwrite. It is obvious that the surrounding layers on the disc are capable of handling the momentary heating sequences.

Read-out of the disc

As there exist two types of disc, the read-out system is capable of two different inputs.

The pre-mastered disc is similar to a compact disc. The same laser is used as during recording, but at lower power level, laser return will vary in intensity, depending on the information on the disc. Figure 19.4 shows a disc read-out.

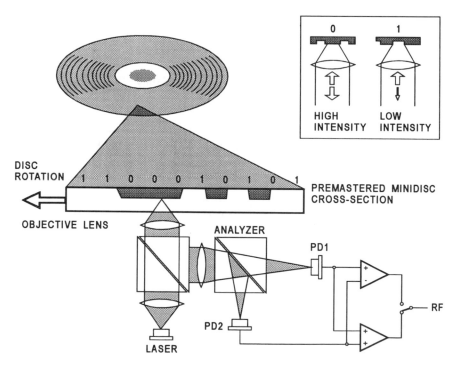

Figure 19.4 *Disc readout*

The recordable disc uses another system, as data is not recorded by a pit system, but in a magnetic layer. Read-out in this case is also performed with the laser.

Laser light hits the disc surface, passes through the magnetic layer and is then reflected back by the reflective layer. However, by passing through the magnetic layer, the polarization of the laser light is changed, according to the magnetic contents of the FeTbCo layer. This phenomenon is called the Kerr effect: polarization of a light beam is influenced by the magnetic medium through which it passes. Figure 19.5 shows an MO read-out.

The Wollaston Principle

For read-out of an MD, there exist two possibilities.

● Read-out of CD-type (pre-mastered) disc: RF is same as CD player
● Read-out of MO-type (recordable) disc: RF is continuous, but with polarization differences.

In order to obtain correct read-out for both possibilities using only one read-out block, a special type of prism was included in the optical block assembly: the Wollaston prism. Figure 19.6 illustrates the Wollaston principle.

A Wollaston prism is a combination of two bonded rock crystals. If a laser light is returned from an MO-type disc, it is sent through this crystal, which has the capacity of giving a multiple output, with main beam (same as

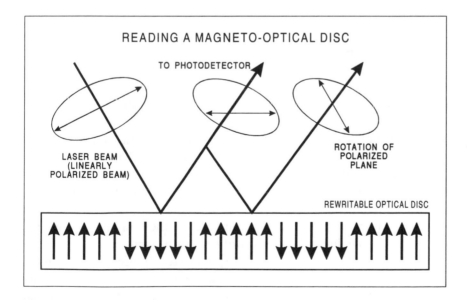

Figure 19.5 *M O readout*

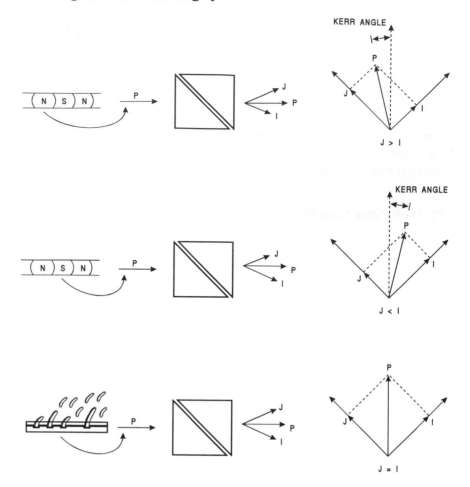

Figure 19.6 *Wollaston principles*

incident beam) and side beams. These side beams (I and J beam in the figure) are directly related to the polarized contents of the incident laser beam.

If the laser light passes through an N magnetized layer on the disc, the polarization of the laser will be influenced by this magnetic content, and one of the side beams (J beam in the figure) will be bigger than the other.

If the laser light passes through an S magnetic layer on the disc, the polarization of the laser will be influenced, but this time the I beam gives a higher output.

As the incident beam (P beam) is composed of the I and the J beam, the Kerr angle, or the direction of polarization, can be seen.

The third possibility is the CD-type (pre-mastered) disc: in this case the laser light does not pass through any magnetic layer, so there will not be any polarization differences and, as a consequence, I and J beams will be equal in level.

MD Detector Block

The detector block is similar to that of a CD player, but I and J detectors are added. Figure 19.7 shows an MD detector block.

In the case of a pre-mastered disc, I and J are added, and as they are equally big, the variations of input will also be equally big and the RF output will appear exactly the same as a CD disc.

In case of a recordable disc, I and J are subtracted. As in this case I and J are not equally big, then depending on the read-out from the disc, the result of this subtracting will be another RF signal which contains all the data information from the disc.

Other signals are similar to those in a CD player, except from the $(A+D)-(B+C)$ signal for address in pre-groove (ADIP) detection.

ADIP

Each recordable (MO) disc has a physical groove, called the address in pre-groove (ADIP). This is a U-shaped depression, stamped into the disc during disc manufacturing. The pre-groove is exactly the same on each disc and follows the spiral data track. However the pre-groove itself shows a specified pattern, as it shows a biphase modulated signal with a basic frequency of 22.05 kHz and modulated to 21.05 and 23.05 kHz. Figure 19.8 shows a pre-groove.

reading out the ADIP information with A, B, C, D detectors; by demodulating the signal, a unique addressing pattern is retrieved which allocates to each position on the disc a correct address. The same address will be used in the recorded data format.

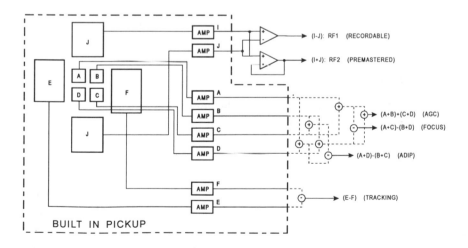

Figure 19.7 *MD detector block*

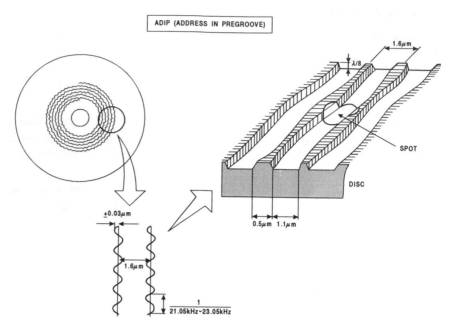

Figure 19.8 *Pre-groove*

Why use ADIP? A recordable disc, when not recorded yet, is totally blank, it contains no information. However, if such a disc did not contain any information at all, especially address information, it would be nearly impossible for any read-out system to perform correct positioning. ADIP serves as a non-erasable addressing possibility and is therefore included.

Read-out of ADIP is similar to a CD read-out because the laser beam lands partly on the disc surface and partly on the depression, which results in a modulated RF.

Track Layout

A pre-mastered disc is also similar to the compact disc with regard to table of contents (TOC), program area and lead-out area.

The recordable disc, on the other hand, has, apart from the lead-in area (containing a TOC) also a user TOC (UTOC) area. UTOC is the area where start and end addresses of music tracks are written.

Note that MD offers the possibilities to change track numbers, to divide tracks and so on. These features will be implemented in the UTOC area. For example, if the user wants to divide one track into two tracks, nothing will happen with the music data in the program area, but the addresses and TOC information in the UTOC will be changed. Figure 19.9 shows a disc track layout.

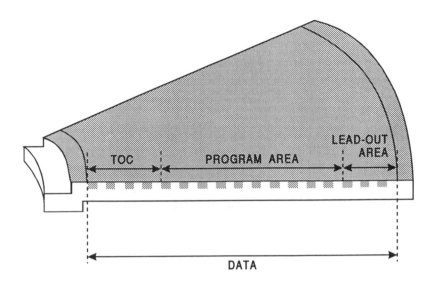

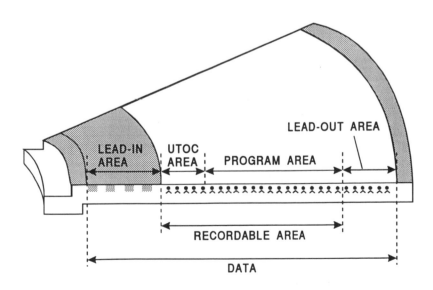

Figure 19.9 *Disc track layout*

Data Format

The data format is similar to the format of a CD, but it also uses parts of the CD-ROM format.

Firstly, when encoding, CIRC is also used, but it is an advanced CIRC format with more interleave (see also Chapter 9) – this is referred to as ACIRC. Figure 19.10 shows the data format.

Secondly, the cluster format was introduced; each cluster contains 36 sectors, the sector format being the same as in a CD-ROM. A cluster contains 32 data sectors and four link sectors. The link sectors depend on the type of disc. In the pre-mastered disc, the four link clusters are fixed subdata sectors which can be used for information on discs, graphics, etc. Figure 19.11 shows the link sectors.

The use of link clusters is needed for the recordable disc. A cluster is the smallest recordable block. It is obvious that between any two of these recordable blocks there must be a buffer area to avoid accidental overwriting.

The first three sectors of each cluster will be used as link, the fourth one for sub-data. This means that the sub-data capacity of a recordable disc is

PLAYBACK-ONLY MINIDISC

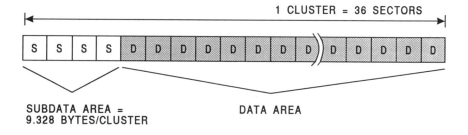

RECORDABLE MINIDISC

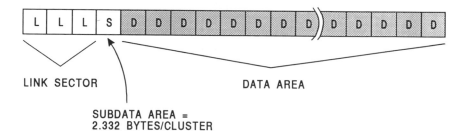

Figure 19.10 *Data format*

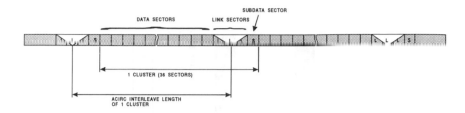

Figure 19.11 *Link sectors*

only a quarter of that of the pre-mastered disc, but remember that this sub-data is redundant (feature) data. Figure 19.12 shows the cluster format. Each sector contains 2352 bytes, of which 2332 bytes are data bytes, the first 20 bytes being sync, cluster/sync address, mode and separator bytes.

A further subgroup is the sound group; each sound group contains 424 bytes, of which 212 bytes are left channel music information and 212 bytes right channel music information. A sector contains 5.5 sound groups, i.e. five full sound groups, and one left or right channel part of a sound group. The corresponding channel bytes will then be in the next sound group.

The Adaptive Transform Acoustic Coding (ATRAC) System

This is a key part of the MD technology. Audio input is encoded/decoded using transform methods which are adaptive and based upon acoustic phenomena.

The aim is to compress the data while maintaining a high degree of audio fidelity.

A normal CD system (16 bit, 44.1 kHz sampling, two channels) has a bit stream of 1.4 Mbit/s. This is before EFM and CIRC processing. On the disc itself it becomes about 1.4 Mbit/s.

In order to put the same amount of music data on a much smaller disc the bit rate has to be reduced. Figure 19.13 shows the compression size.

It is obvious that a reduction in the bit rate affects the audio quality and the system's software has to ensure that audio quality is high enough to satisfy the expectations of today's sophisticated listeners.

Input in the ATRAC system is the same 16 bit/two channel/44.1 kHz bit stream. Output, however, is about 292 kbit/s.

Psycho-acoustics

The reduction of the bit rate without altering the audio quality is based on some psycho-acoustic principles.

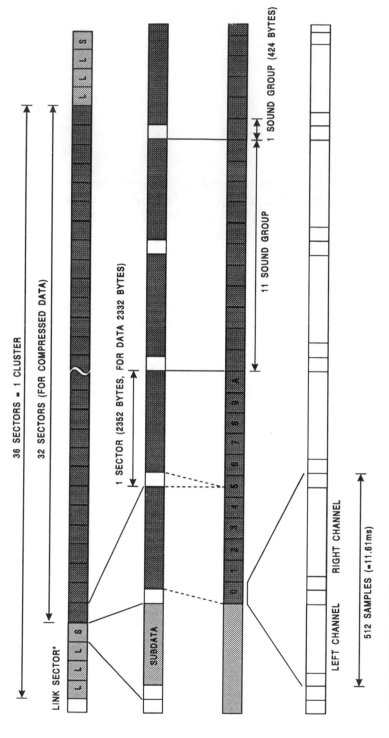

Figure 19.12 *Cluster format*

36 SECTORS = 1 CLUSTER

32 SECTORS (FOR COMPRESSED DATA)

LINK SECTOR*

1 SECTOR (2352 BYTES, FOR DATA 2332 BYTES)

11 SOUND GROUP

1 SOUND GROUP (424 BYTES)

SUBDATA

LEFT CHANNEL   RIGHT CHANNEL

512 SAMPLES (=11.61ms)

* THIS BECOMES A SUBDATA SECTOR IN PLAYBACK-ONLY MINIDISCS.

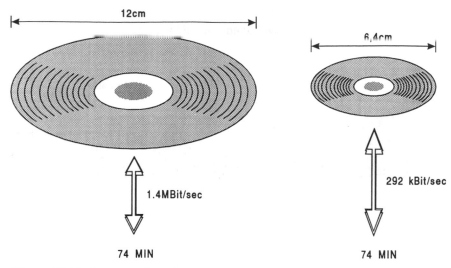

Figure 19.13 *Compression size*

The sensitivity of the human ear varies according to the frequency range. A sound, at a given level and frequency, can be perceived by our hearing system (ears and brain), while another sound, at an even higher level, but at another frequency might not be perceived. Figure 19.14 illustrates the psycho-acoustic affects and thresholds. It is therefore possible to define over the total frequency spectrum so-called hearing thresholds.

An analysis of the input music allows us to define which parts are below this threshold; these can then be deleted and the bit rate can thus be reduced.

The sensitivity of our hearing system is equal in certain so-called critical bands. For instance, if we wanted to test the sensitivity at around 100 Hz, we would find that within a bandwidth of 100 Hz of this frequency the sensitivity is the same. The higher the frequency range, the bigger the bandwidth of the critical bands (same sensivity level). At 100 Hz this bandwidth is about 160 Hz and at 10 000 Hz it is 2500 Hz.

About 25 critical bands are defined. The use of these critical bands in audio analysis allows the reduction of the bit rate. However, in order to keep as closely as possible to the original input, ATRAC uses more critical bands (about 52).

Another important fact is the masking effect. If two sound sources emit sound at the same time, but at different levels, it is very likely that one of the two sound sources will mask the other as its level is much higher. For example, if an airplane passes overhead while two people are having a conversation, for a certain time the conversation will be masked by the much higher level of the passing airplane. Figure 19.5 illustrates this masking effect.

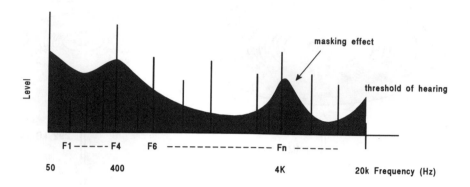

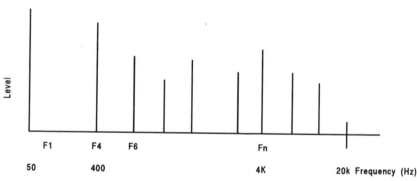

Figure 19.14 *Psychoacoustic effects : threshold*

By analysing a certain amount of input samples, it is possible to define which parts of the input frequencies are masking other frequencies. The masked frequencies can then be deleted and the bit rate can thus be reduced.

ATRAC Operation

The psycho-acoustic basics explained above will be used, in the correct sequence, to reduce the 1.4 Mbit/s input to 292 kbit/sec.

However, it should be noted from the start that the ATRAC system allows the change of the operational software, i.e. the used calculation algorithms, without creating compatibility problems.

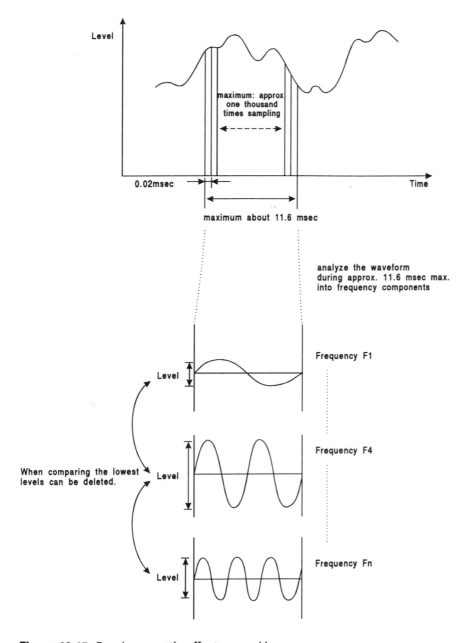

Figure 19.15 *Psychoacoustic effects : masking*

The input audio (16 bit/two channel/44.1 kHz rate) is firstly analysed for frequency bandpassing. Based upon this time domain analysis, a first bandpassing is performed where the high frequencies (11.05–22.05 kHz) are split; in a second stage, the mid-range and lower range (5.5025–11.025 kHz, 0–5.5025 kHz respectively) are split while the high frequencies are delayed to keep correct timing with the other bands while they are processed. Figure 19.16 presents an encoding block diagram.

Then follows the transition from the time domain to the frequency domain, based upon modified discrete cosine transforms (MDCT) algorithms. In fact, this analysis is similar to the Fourier analysis where an input spectrum is analysed to determine its frequencies (and their respective levels).

Before the transition from the time domain to the frequency domain, a block decision needs to be taken. This is the amount of time/frequency blocks that will be used to analyse the input sound.

As mentioned before, a number of critical frequency bands are defined. ATRAC uses a lot more critical bands than the minimum needed and it is therefore possible to assess the exact number of bands from timeblock to timeblock. However, the timing can also be adapted, i.e. the number of samples that will be analysed as one block: the maximum is 11.6 ms (or 512 samples at 44.1 kHz).

The reason why this time/frequency block flexibility is included is also based upon psycho-acoustics. When we hear a short, impulse-type sound, we perceive mainly big level differences rather than the more subtle ones. When on the other hand we listen to a slow changing sound, we are more likely to perceive subtle differences because there are no big level differences. The time/frequency block decision allows ATRAC to perform this task: depending on the input sound (big or small level differences compared with input to input block) more or less time and the correct number of frequency bands are allocated.

In doing so, the bit rate can also be reduced because any meaningless information is deleted. Figure 19.17 shows the ATRAC encoding process.

When this block decision is performed, on each block there will be MDCT calculations which, on a mathematical basis, analyse the input sound. Once this task is performed, the psycho-acoustic principles can be applied also on a mathematical basis.

Finally, bit allocation is performed. Each frequency/time block is analysed, along with its level. This is not represented by a fixed amount of bits, but by a variable amount, from 0 to 15, representing the dynamic range of that part. Along with the dynamic range, a scale factor gives the relative level of the signal, and the remaining bits can be deleted as this is quantization noise. In this way, only the significant part of the data is encoded. Figure 19.18 shows ATRAC data.

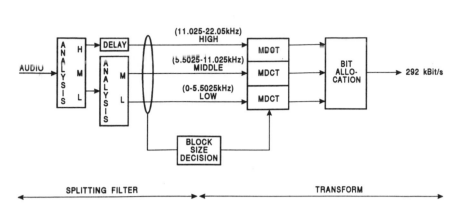

Figure 19.16 *Encoding block diagram*

As a result of this ATRAC encoding, however, the output data should not be described as audio data, but as data describing parts of the audio spectrum in word lengths and scale factors, comparable with floating block data in computers.

Output of ATRAC encoding will be EFM and ACIRC processed, similar to CD encoding.

ATRAC Read-out

When data is read out from an MD, it is first EFM and ACIRC decoded (again similar to a CD) and the obtained audio data (word lengths, scale factors, spectrum data) are then used to perform an inversed ATRAC calculation which will reconstruct the original input data. Figure 19.19 shows the ATRAC decoding process.

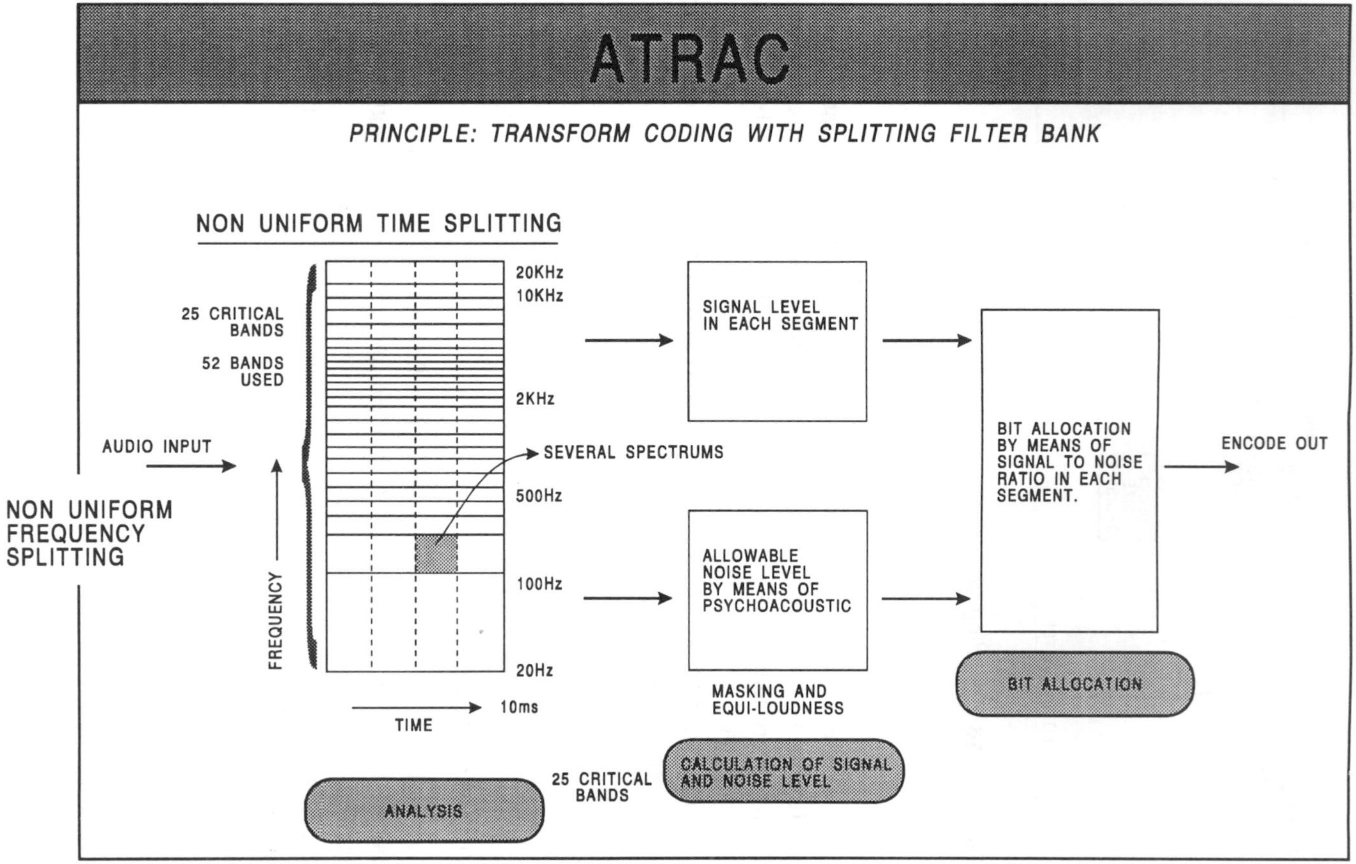

Figure 19.17 ATRAC encoding

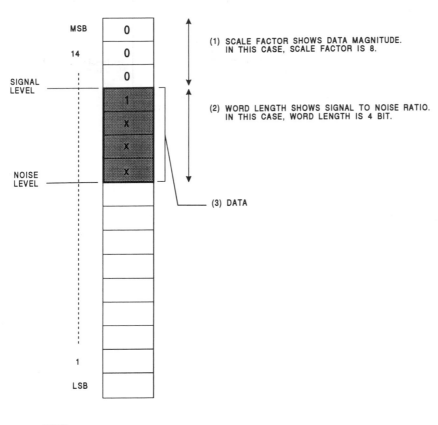

(1) SCALE FACTOR SHOWS DATA MAGNITUDE.
IN THIS CASE, SCALE FACTOR IS 8.

(2) WORD LENGTH SHOWS SIGNAL TO NOISE RATIO.
IN THIS CASE, WORD LENGTH IS 4 BIT.

(3) DATA

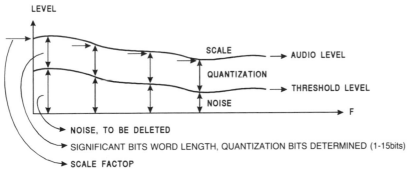

NOISE, TO BE DELETED

SIGNIFICANT BITS WORD LENGTH, QUANTIZATION BITS DETERMINED (1-15bits)

SCALE FACTOP

Figure 19.18 *ATRAC data*

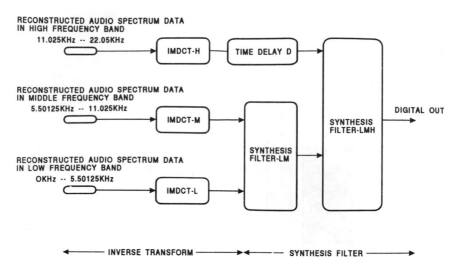

Figure 19.19 *ATRAC decoding*

20
Digital Compact Cassette (DCC) Format

The Digital Compact Cassette (DCC) can be considered as an update of the very popular Analog Compact Cassette (CC). The design objectives were to create a successor for the CC that would be backwards compatible, meaning that a DCC player can also playback CCs. Tape width (3.8 mm) and tape speed (4.76 cm/s) remain equal to achieve this compatibility. An advantage for the manufacturer of DCC equipment is that existing tape transport mechanisms can be used, thereby reducing development costs.

Dimensions of the DCC cassette are basically the same as those of the CC. The most obvious difference is the addition of a slider to protect the tape against dust and damage, similar to the cover found on video cassettes. When closed the slider also locks the reels. Although the tape is recordable on two sides, the reels are exposed on only one side. The closed side of the cassette can be used for artwork and/or a list of music titles. As a consequence, all DCC machines are equipped with an autoreverse function.

Figure 20.1 shows the block diagram of a DCC recorder. Three main parts can be identified: the input/output interface, the precision adaptive subband coding (PASC) processor and the tape recording and playback system. In recording mode, the digital input signal from the input interface is processed by the PASC encoder to reduce the amount of source code by about four times. The encoded data is formatted into frames and recorded onto eight parallel tape tracks. Eight to ten modulation is used to reduce low frequency components in the recorded signals. Error-correcting code and synchronization signals are added in the tape drive section. Auxiliary code, such as track numbers, time information and table of contents are recorded at a much lower bit rate on a separate tape track.

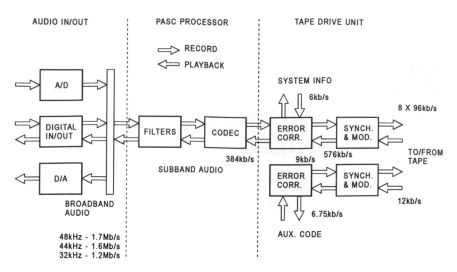

Figure 20.1 *Block diagram of DCC recorder*

By using eight parallel tracks, the recorded bit rate could be reduced to 96 kbit/s/track, the corresponding minimum wavelength equals:

$$\lambda = \frac{v}{f} = \frac{4.76 \text{ cm/s}}{96 \text{ kbit}} \times 2 \text{ bit/Hz} = 0.99 \, \mu\text{m}$$

Normal video tape can cope with a minimum wavelength of about 1 μm. The DCC tapes therefore do not have to rely on special tape formulations, instead, standard video tape that is produced in high volumes can be used.

Head Layout

Figure 20.2 represents schematically the head layout of a DCC recorder. The assembly combines three sets of thin film heads: one set of nine integrated recording heads (IRH) for digital recording, one set of nine magneto-resistive heads (MRH) for digital playback and one set of two MRHs for analog playback. Digital signals are recorded on nine parallel tracks, each 185 μm wide with a track pitch of 195 μm. The digital playback heads are only 70 μm wide to guarantee interchangeability of cassettes between different players. It also makes the DCC format less sensitive to azimuth errors than the CC. Note that a DCC tape always has its signals on the opposite half of the tape compared with a CC recording in the same direction.

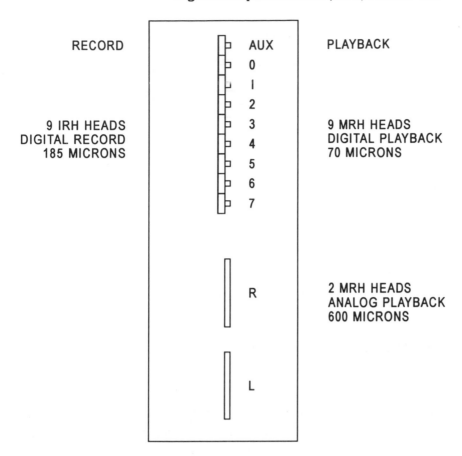

RECORD

9 IRH HEADS
DIGITAL RECORD
185 MICRONS

AUX
0
I
2
3
4
5
6
7

R

L

PLAYBACK

9 MRH HEADS
DIGITAL PLAYBACK
70 MICRONS

2 MRH HEADS
ANALOG PLAYBACK
600 MICRONS

Figure 20.2 *DCC head layout*

Data Format

Eight parallel tracks carry the main data (PASC and system information).
Fig. 20.3 shows the track format for the main data channels. The signals on
tape are divided into tape frames, separated by a variable length inter frame
gap (IFG). The purpose of this IFG is to compensate for tolerances in the
sampling frequency (e.g. clock jitter) during recording. The IFG consists of
a signal with alternating polarity at every bit position.

Each of the eight tape frames contains 32 tape blocks of 51 tape symbols.
A symbol is a 10-bit word generated from the original 8-bit data by 8–10
modulation. The first three symbols in every block form a tape header that
includes a synchronization pattern and a frame and block address. The
synchronization pattern serves to identify the start of a tape block while
the addresses are needed to reconstruct the data during playback. The
remaining 48 symbols, called the body, carry PASC audio data, system

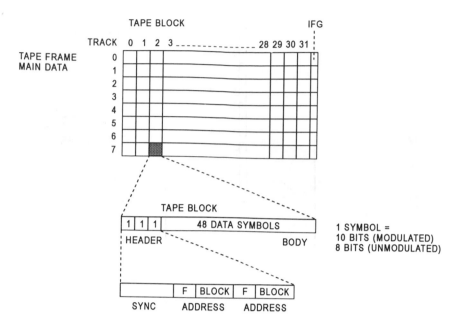

Figure 20.3 *Main data track format*

information and error-correcting code. A CIRC is employed to protect the data against random and burst errors. A main data tape frame contains $48 \times 32 = 1632$ symbols (or bytes) per track or 13,056 symbols in total. When excluding the header, 12,288 bytes remain of which there are 8192 bytes for the PASC audio data, 3968 parity bytes for error correction and 128 bytes of system information.

One track is available for auxiliary data. The bit rate of the auxiliary track is reduced to one eighth of the bit rate of the main data tracks (12 kbit/s). Hence, the number of tape blocks reduces to four (Fig. 20.4).

Precision adaptation sub-band coding (PASC)

The purpose of the coding algorithm is to reduce the original bit rate of the digital audio signal by about four times to conform with the 384 kbit/s maximum bit rate of the DCC system. PASC relies on two psycho-acoustic principles of the human ear, similar to the ATRAC system used in MD.

The human ear can only hear sounds above a certain level, called the hearing threshold. This depends on the frequency of the sound as shown in Fig. 20.5. In addition, loud sounds mask adjacent softer sounds by raising the hearing threshold as demonstrated in Fig. 20.6. By calculating the

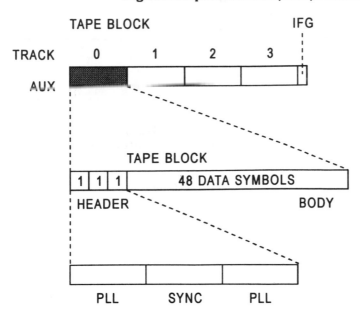

Figure 20.4 *Auxiliary data track format*

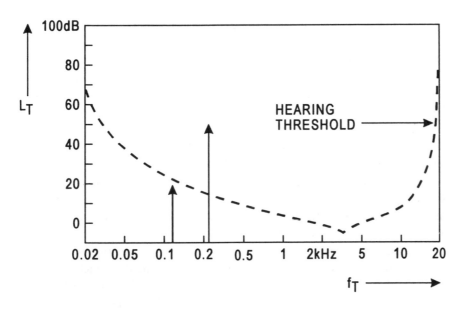

Figure 20.5 *Absolute hearing threshold*

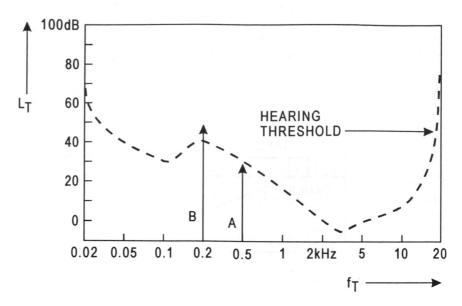

Figure 20.6 *Loud signals (B) mask soft signals (A) by raising the hearing threshold*

hearing threshold for all incoming samples and only transferring signal components that extend above this threshold, a very efficient coding of audio signals can be achieved.

Another psycho-acoustic principle states that the human ear perceives sounds by using critical bands. It analyses sounds by small frequency bands centred around the frequency of the sound. The bandwidth of a critical band increases with the frequency from approximately 100 Hz wide below 500 Hz to 2500 Hz wide at 10 kHz. As a result, any psycho-acoustic coding algorithm must perform its calculations by taking into account the critical bands. This leads to sub-band filters that divide the frequency spectrum into unequal bandwidth sub-bands. In the PASC system, however, sub-band filters with equal bandwidth sub-bands are used. This has the advantage of reduced filter complexity. To compensate for the equal bandwidth sub-bands a different bit allocation algorithm is used. In a critical bandwidth system the number of bits allocated to each sub-band will, on average, be equal as shown in Fig. 20.7. With the PASC equal bandwidth system, the number of bits allocated will be higher for the lower frequency sub-bands and lower for the higher frequency sub-bands.

The PASC processor (Fig. 20.8) analyses the broadband audio signal with sampling frequency F_s by dividing it into 32 sub-band samples with sampling frequency $F_s/32$. Filtering is performed by using a window representing a 512 tap FIR filter, as shown in Fig. 20.9. The window is shifted 32 samples for every calculation of 32 new sub-band samples (one sample for

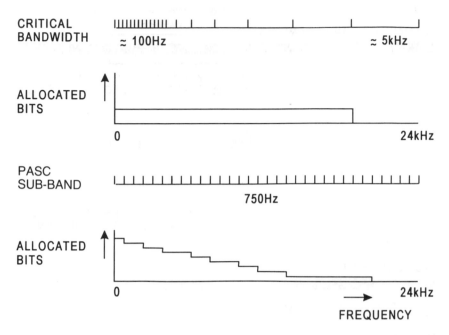

Figure 20.7 *difference in bit allocation between systems based on critical bandwidths and the PASC system with equal bandwidth sub-bands*

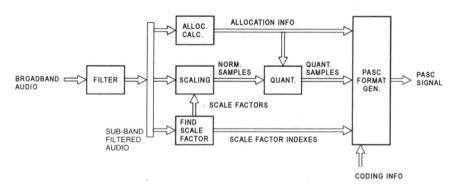

Figure 20.8 *Block diagram of the PASC encoder*

each sub-band). Because of the large overlap any sub-band signal will still be a PCM signal, only band limited to the sub-band. Sub-band signals are represented in fixed point notation. PASC signals use a floating point notation. The length of the mantissa determines the resolution of the system, which is at maximum 15 bits (corresponding to a theoretical S/N ratio of 92 dB). The exponent controls the dynamic range, it is chosen to cover a range of $+6$ dBFS to -118 dBFS in 2 dB steps.

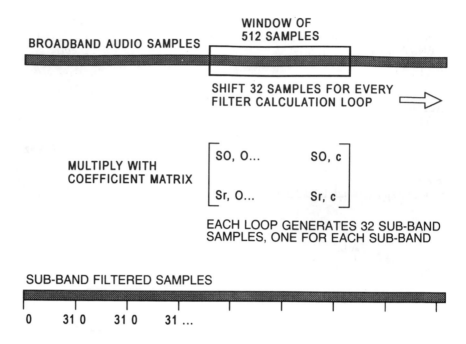

BROADBAND AUDIO SAMPLES

WINDOW OF
512 SAMPLES

SHIFT 32 SAMPLES FOR EVERY
FILTER CALCULATION LOOP

MULTIPLY WITH
COEFFICIENT MATRIX

$$\begin{bmatrix} S0, 0... & S0, c \\ Sr, 0... & Sr, c \end{bmatrix}$$

EACH LOOP GENERATES 32 SUB-BAND
SAMPLES, ONE FOR EACH SUB-BAND

SUB-BAND FILTERED SAMPLES

0   31 0   31 0   31 ...

Figure 20.9 *Filtering of broadband audio into sub-band signals*

The masking principles can be implemented in the representation of the sub-band samples by varying the length of the mantissa, the resolution must only be precise enough to keep the quantization noise below the masking threshold.

A PASC signal is composed of PASC frames. Each frame contains information on 12 sub-band samples or 384 samples in total. The frame period equals $F_s/384$ (8 ms for $F_s = 48\,kHz$).

In order to decide the masking threshold, first the power of the sub-band signals for the corresponding PASC frame period is calculated. These are then used to find the contribution of the masking between sub-bands or the masking within the sub-band itself. Then the absolute hearing threshold is applied. The resulting masking threshold is compared with the peak powers of the sub-band signals. Two situations are possible.

● Peak power of a sub-band is lower than the masking threshold. These sub-bands contain no relevant information and must therefore not be transferred

● Peak power of a sub-band is higher than the masking threshold. These sub-band signals will be transferred with a mantissa the length of which is proportional to the distance between the peak power and the masking threshold.

Because the masking threshold is recalculated for every PASC frame transferred, the process is called adaptive allocation of the available capacity.

When the length of the mantissa has been determined the conversion of the PCM sub-band samples to floating point notation is performed. For every sub-band the absolute value of the peak value of 12 samples is compared to a table of multiplication constants (so-called scale factors) that represent the 2 dB steps in dynamic range. The samples of the sub-band are normalized onto the scale factor by division. The scale factor represents the exponent of the floating point notation; it is the same for all the coded samples of the sub-band and will be recorded along with the coded samples.

After normalization the samples represent the mantissa. These are now quantized, depending on the number of bits allocated to the sub-band; if, e.g., only 4 bits are allocated a 15 level quantization is performed.

In the PASC decoder (Fig. 20.10) encoded samples are reconverted to PCM sub-band samples by multiplication with the scale factor. The synthesis filter merges the sub-bands into a broadband audio signal. For correct decoding the PASC frame must contain the following information.

● A synchronization pattern
● Coding information such as the original sampling frequency, emphasis, etc.
● The allocation information for the 32 sub-bands, included as 4-bit values that tell the decoder how many bits were used for the coding of the mantisssa. If an entry has value 0 it means peak power of the sub-band was below the masking threshold and thus the sub-band was not transferred
● The scale factors for each of the sub-bands, represented as 6-bit values
● The PASC encoded samples
● Additional 0 bits to adjust the PASC frame rate to Fs/384.

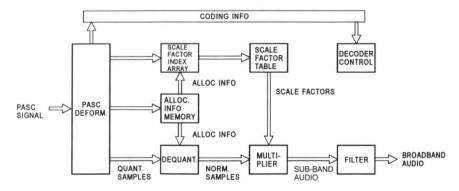

Figure 20.10 *Block diagram of the PASC decoder*

Appendix 1

Error Correction

Error correction is one of the most advanced areas in the entire field of digital audio. It is purely because of error-correction techniques that reliable digital recordings can be made, despite the frequent occurrence of tape dropouts.

In the next two sections, we are going to discuss the theory behind the EIAJ format P, Q and CRC codes. Because of the special nature and the complexity involved, a mathematical treatment of this topic is unavoidable. However, every attempt has been made to pull all material together in a concise and systematic manner. Examples are included to help you see the situations more vividly. For your part, the only prerequisite in reading them is some knowledge of matrix algebra.

P, Q and the cyclic redundancy check code

In Figure A1.1, the P codes are generated by feeding the input data $L_0 - R_2$ to an exclusive-or gate. Hence, we have:

$$P_0 = L_0 \oplus R_0 \oplus L_1 \oplus R_1 \oplus L_2 \oplus R_2 \ldots \qquad (1)$$
$$P_1 = L_3 \oplus R_3 \oplus L_4 \oplus R_4 \oplus L_5 \oplus R_5$$
$$P_n = L_{3n} \oplus R_{3n} \oplus L_{3n+1} \oplus R_{3n+1} \oplus L_{3n+2} \oplus R_{3n+2}$$

The symbol $\oplus$ (pronounced ring-sum) indicates **modulo-2 summation**, which obeys the following rules:

$$0 \oplus 0 = 0$$
$$0 \oplus 1 = 1$$
$$1 \oplus 0 = 1$$
$$1 \oplus 1 = 0$$

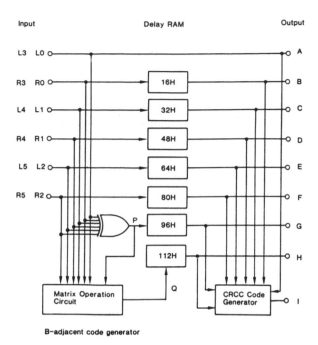

Figure A1.1

$$Q_0 = T^6L_0 + T^5R_0 + T^4L_1 + T^3R_1 + T^2L_2 + TR_2$$

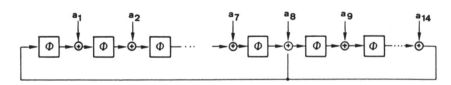

Figure A1.2

which is an **exclusive-or** operation. When considering expression (1) and applying the rules, the parity bit P_0 will be 0 for even numbers of logic levels, and 1 for odd numbers.

Q codes are generated by a **matrix operation circuit**, the principle of which is shown in Figure A1.2.

Initially, L_0, a 14-bit data word comprising bits a_1 to a_{14}, is applied to the summing nodes and the shift register contents are all 0s. After one shift to the right, the shift register contents change to those shown in Figure A1.3. The single shift operation on word L_0 can be defined as TL_0.

Now, a second data word R_0, comprising bits b_1 to b_{14}, is applied to the summing nodes. When the next shift occurs, the shift register contents

1Shift

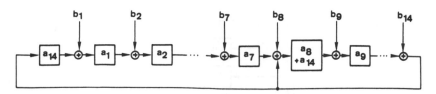

Figure A1.3

2 Shift

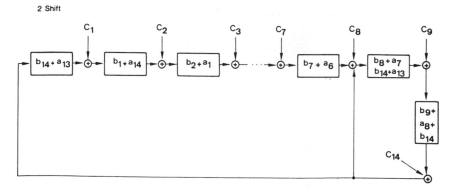

Figure A1.4

change to those shown in Figure A1.4. This time, L_0 undergoes two shifts while R_0 is subjected to only one, the combined effect can be defined as $TTL_0 \oplus TR_0$, or $T^2 L_0 \oplus TR_0$.

After six shift operations, the shift register contents constitute the code word Q_0. Mathematically we can write:

$$Q_0 = T^6 L_0 \oplus T^5 R_0 \oplus T^4 L_1 \oplus T^3 R_1 \oplus T^2 L_2 \oplus TR_2$$

Similarly we have,

$$Q_1 = T^6 L_3 \oplus T^5 R_3 \oplus T^4 L_4 \oplus T^3 R_4 \oplus T^2 L_5 \oplus TR_5$$

and

$$Q_n = T^6 L_{3n} \oplus T^5 R_{3n} \oplus \ldots \oplus T^2 L_{3n+2} \oplus TR_{3n+2}$$

The 1-bit shift and 2-bit modulo-3 summation are functions of T having the following forms:

$$T = \begin{bmatrix} 0 & 0 & 0 & 0 & 0 & 0 & 0 & 0 & 0 & 0 & 0 & 0 & 0 & 1 \\ 1 & 0 & 0 & 0 & 0 & 0 & 0 & 0 & 0 & 0 & 0 & 0 & 0 & 0 \\ 0 & 1 & 0 & 0 & 0 & 0 & 0 & 0 & 0 & 0 & 0 & 0 & 0 & 0 \\ 0 & 0 & 1 & 0 & 0 & 0 & 0 & 0 & 0 & 0 & 0 & 0 & 0 & 0 \\ 0 & 0 & 0 & 1 & 0 & 0 & 0 & 0 & 0 & 0 & 0 & 0 & 0 & 0 \\ 0 & 0 & 0 & 0 & 1 & 0 & 0 & 0 & 0 & 0 & 0 & 0 & 0 & 0 \\ 0 & 0 & 0 & 0 & 0 & 1 & 0 & 0 & 0 & 0 & 0 & 0 & 0 & 0 \\ 0 & 0 & 0 & 0 & 0 & 0 & 1 & 0 & 0 & 0 & 0 & 0 & 0 & 0 \\ 0 & 0 & 0 & 0 & 0 & 0 & 0 & 1 & 0 & 0 & 0 & 0 & 0 & 1 \\ 0 & 0 & 0 & 0 & 0 & 0 & 0 & 0 & 1 & 0 & 0 & 0 & 0 & 0 \\ 0 & 0 & 0 & 0 & 0 & 0 & 0 & 0 & 0 & 1 & 0 & 0 & 0 & 0 \\ 0 & 0 & 0 & 0 & 0 & 0 & 0 & 0 & 0 & 0 & 1 & 0 & 0 & 0 \\ 0 & 0 & 0 & 0 & 0 & 0 & 0 & 0 & 0 & 0 & 0 & 1 & 0 & 0 \\ 0 & 0 & 0 & 0 & 0 & 0 & 0 & 0 & 0 & 0 & 0 & 0 & 1 & 0 \end{bmatrix}$$

and

$$TL_0 = T \begin{bmatrix} a_1 \\ a_2 \\ a_3 \\ a_4 \\ a_5 \\ a_6 \\ a_7 \\ a_8 \\ a_9 \\ a_{10} \\ a_{11} \\ a_{12} \\ a_{13} \\ a_{14} \end{bmatrix} = \begin{bmatrix} a_{14} \\ a_1 \\ a_2 \\ a_3 \\ a_4 \\ a_5 \\ a_6 \\ a_7 \\ a_8 \oplus a_{14} \\ a_9 \\ a_{10} \\ a_{11} \\ a_{12} \\ a_{13} \end{bmatrix}$$

If we let:

$$\begin{bmatrix} L_0 = a_1 a_2 \ldots a_{14} \end{bmatrix}$$

$$\begin{bmatrix} R_0 = b_1 b_2 \ldots b_{14} \end{bmatrix}$$

$$\begin{bmatrix} L_1 = c_1 c_2 \ldots c_{14} \end{bmatrix}$$

.

.

.

$$\begin{bmatrix} R_2 = f_1 f_2 \ldots f_{14} \end{bmatrix}$$

Then we have the following:

$$TR_2 = \begin{bmatrix} f_{14} \\ f_1 \\ f_2 \\ f_3 \\ f_4 \\ f_5 \\ f_6 \\ f_7 \\ f_8 \oplus f_{14} \\ f_9 \\ f_{10} \\ f_{11} \\ f_{12} \\ f_{13} \end{bmatrix} \quad T^2 L_2 = \begin{bmatrix} e_{13} \\ e_{14} \\ e_1 \\ e_2 \\ e_3 \\ e_4 \\ e_5 \\ e_6 \\ e_7 \oplus e_{13} \\ e_8 \oplus e_{14} \\ e_9 \\ e_{10} \\ e_{11} \\ e_{12} \end{bmatrix} \quad T^3 R_1 = \begin{bmatrix} d_{12} \\ d_{13} \\ d_{14} \\ d_1 \\ d_2 \\ d_3 \\ d_4 \\ d_5 \\ d_6 \oplus d_{12} \\ d_7 \oplus d_{13} \\ d_8 \oplus d_{14} \\ d_9 \\ d_{10} \\ d_{11} \end{bmatrix}$$

$$T^4 L_2 = \begin{bmatrix} c_{11} \\ c_{12} \\ c_{13} \\ c_{14} \\ c_1 \\ c_2 \\ c_3 \\ c_4 \\ c_5 \oplus c_{11} \\ c_6 \oplus c_{12} \\ c_7 \oplus c_{13} \\ c_8 \oplus c_{14} \\ c_9 \\ c_{10} \end{bmatrix} \quad T^5 R_0 = \begin{bmatrix} b_{10} \\ b_{11} \\ b_{12} \\ b_{13} \\ b_{14} \\ b_1 \\ b_2 \\ b_3 \\ b_4 \oplus b_{10} \\ b_5 \oplus b_{11} \\ b_6 \oplus b_{12} \\ b_7 \oplus b_{13} \\ b_8 \oplus b_{14} \\ b_9 \end{bmatrix} \quad T^6 L_0 = \begin{bmatrix} a_9 \\ a_{10} \\ a_{11} \\ a_{12} \\ a_{13} \\ a_{14} \\ a_1 \\ a_2 \\ a_3 \oplus a_9 \\ a_4 \oplus a_{10} \\ a_5 \oplus a_{11} \\ a_6 \oplus a_{12} \\ a_7 \oplus a_{13} \\ a_8 \oplus a_{14} \end{bmatrix}$$

Now, let us see how a Q code can be generated if data are input serially. Consider the following scheme:

(1) One of the feedback loops is controlled by a switch that turns on whenever 15 shifts are made.

(2) The exclusive-or operation is only active during a SR (shift register) shift.

With the following configuration:

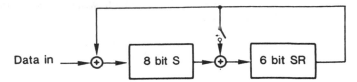

Figure A1.5

Then, after 14 shifts the contents of the two SRs are:

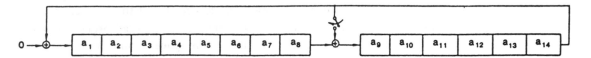

Figure A1.6a

For an additional shift we have,

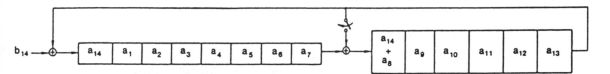

Figure A1.6b

which we recognize as TL_0. As we continue we have,

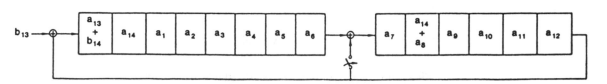

Figure A1.6c

After the 30th shift, the contents of the SRs are:

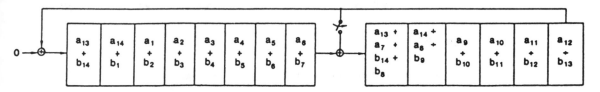

Figure A1.6d

which is $T^2 L_0 + TR_0$ (when compared with the two column matrixes TR_2 and $T^2 L_2$). Note that for this scheme to work, the serial data have to be sent in the following format:

$$a_{14}\, a_{13}\, a_{12}\, a_{11} \ldots a_1 \qquad bb_{14}\, b_{13}\, b_1 \qquad bc_{14}\, c_{13}$$

where $b = 0$. Finally, the reader can easily verify that after 90 shifts, the contents of the two SRs constitute the word Q.

The PCM-F1 employs cyclic redundancy check code (CRCC) for **detecting** code errors. The encoding can be easily implemented by using shift registers while the decoding scheme becomes simple because of the inherent well-defined mathematical structure. In mathematical terms, a **code word** is a **code vector** because it is an n-tuple from the **vector space** of all n-tuples. For a linear code C with length n and containing r information digits, if an n-tuple

$$V^{(1)} = (V_{n-1}, V_1, \ldots, V_{n-2})$$

obtained by shifting the code vector of C

$$V = (V_0, V_1, V_2, \ldots V_{n-1})$$

cyclically one place to the right is also a code vector of C, then linear code C is called a cyclic code. From this definition, it follows that no matter how many times the digits in V are shifted cyclically to the right, the resulting n-tuple is also a code vector. The components of a code vector define the coefficients of a polynomial. Therefore, if we denote V (X) the code polynomial of V, we then have:

$$V(X) = V_0 + V_1 X + V_2 X^2 + \ldots + V_{n-1} X^{n-1}$$

or, in a particular case, where we have the following code word,

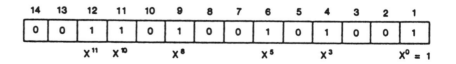

Figure A1.7

The code polynomial is written as,

$$V(X) = X^{11} + X^{10} + X^8 + X^5 + X^3 + 1$$

where + stands for modulo-2 summation. Moreover, every code polynomial V (x) can be expressed in the following form:

$$V(X) = M(X)g(X)\ldots\ldots \qquad (2)$$
$$= (M_0 + M_1 X + M_2 X^2 + \ldots + M_{r-1} X^{r-1})(1 + g_1 X + g_2 X^2 + \ldots + g_{n-r-1} X^{n-r-1} + X^{n-r})$$

where $M_0, M_1, \ldots,_{r-1}$ are the information digits to be encoded and $g(X)$ is defined as a **generator polynomial**. Hence, the encoding of a message $M(X)$ is equivalent to multiplying it by $g(X)$. Note that in any particular cyclic code, there only exists one generator polynomial $g(X)$. However, if we encode the information digits according to (2), the orders of the message digits are altered. Hence, a different scheme must be used. The code can be put into a systematic form (see example 3 below) by first multiplying $M(X)$ by X^{n-r} and then divide the result by $g(X)$, hence,

$$X^{n-r} M(X) = Q(X)g(X) + R(X)$$

Adding $R(X)$ to both sides, we have

$$X^{n-r} M(X) + R(X) = Q(X)g(X)\ldots\ldots \qquad (3)$$

Where $Q(X)$ is the quotient.
$\quad\quad\quad$ $R(X)$ is the remainder.

And, in modulo-2 summation $R(X) + R(X) = 0$

If we define the left-hand side of (3) as our transmission polynomial $T(X)$, then

$$T(X) = Q(X)g(X)$$

which means that $T(X)$ can be **divided exactly** by $g(X)$. In this case, the encoding is equivalent to converting the information polynomial $M(X)$ into $T(X)$.

If, in the recording process, $T(X)$ changes to $E(X)$ and errors are detected. In the PCM-F1, when the CRCC indicates that one word in a horizontal scanning period H is incorrect, the remaining words in that period will all be considered erroneous.

At this point it is natural to ask what kind of properties $g(X)$ must have in order to generate an (n, r) cyclic code, i.e., generating an n-bit code from a message of r bits. We have the following theorem.

Theorem. If $g(X)$ is a polynomial of degree n-r and is a factor of $X^n + 1$, then $g(X)$ generates an (n, r) cyclic code.

Now, we shall show you how a 4-bit cyclic code can be generated from a message of 3 bits. We hope that the whole situation will be clarified by the following simple examples.

Example 1. Since $n = 4$ and $r = 3$, we are searching for a polynomial of degree 1. Because:

$$X^4 + 1 = (X + 1)(X^3 + X^2 + X + 1)$$

Our $g(X)$ will be $X + 1$

Note that $g(X)$ is primitive (irreducible).

Example 2. Consider the messages 000, 100, 010, ... 111 where the LSB is at the extreme left of each word. From (2), we have

for 100, $V(X) = 1(1 + X) = 1 + X$
$\quad\ 101$, $V(X) = (1 + X^2)(1 + X) = 1 + X + X^2 + X^3$

For these two cases, the coded words are 1100 and 1111 respectively.
Proceeding as before, we have the following:

MESSAGES			CODE			
LSB		MSB	LSB			MSB
0	0	0	0	0	0	0
1	0	0	1	1	0	0
0	1	0	0	1	1	0
1	1	0	1	0	1	0
0	0	1	0	0	1	1
1	0	1	1	1	1	1
0	1	1	0	1	0	1
1	1	1	1	0	0	1

This code has a minimum distance of 2, hence it can detect a single error in the message, it is not systematic, i.e., we do not have a code where the first 3 digits are the unaltered message digits and the last one is a check bit.

It is not difficult to see that this code can be implemented by the circuit shown below:

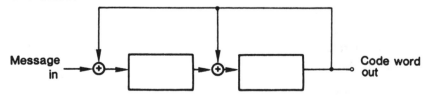

Figure A1.8

which employs two exclusive-or gates and two SRs. For this scheme to work, the message should be input as:

$M_1 M_2 M_3 M_1 0 M_1' M_2' M_3' M_1' 0 \ldots$ etc.

The SRs reset whenever 5 shifts are made.
   The decoder has the following configuration:

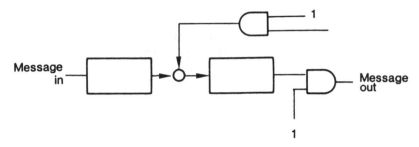

Figure A1.9

After 4 data bits come out of the AND gate, we look at the 4th bit. If this bit is zero, the three message bits are considered correct; otherwise, an error is made. Hence, the SRs reset after every 4 shifts. Also, the message input sequence is:

$C_1 C_2 C_3 C_4 C_1' C_2' C_3' C_4' \ldots$ etc.

The circuit works because when we multiply $a_1 a_2 a_3$ by 11 we get the following:

| a_1 | a_2 | a_3 |
	1	1
a_1	a_2	a_3
	a_1	a_2	a_3
a_1	$a_1 + a_2$	$a_2 + a_3$

Hence, with the following circuit

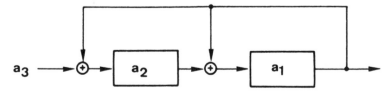

Figure A1.10

2nd shift, a_1 is out.

From these 5 shifts, we get $0, a_1, a_1 + a_{2,}, a_2 + a_3$; which matches the elements generated by the multiplication.

However, for this to work, the message input must have the following format:

$$M_1M_2M_3M_10M_1'M_3'M_1'0\ldots$$

and the SRs reset whenever 5 shifts are made.

Hence, for the decoded message $D_1D_2D_3$, we have:

$$D_1 = a_1.1$$
$$D_2 = ((a_1.1) + a_2).1$$
$$D_3 = (((a_1.1) + a_2).1 + a_3).1$$

Also, $R = D_3 + a_4$

For zero error detection, D_3 equals a_4. This is true when we look at the LSB columns of the message and code table. It is also clear that the decoder circuit implements the above division.

Example 3. We can form a systematic code by using equation (3). Take the last message word (1110 in the above example), then, the message polynomial is $M(X) = 1 + X + X^2$.

Since $X^{n-r} = X$ in this case, we have $XM(X) = X + X^2 + X^3$.

Dividing $XM(X)$ by the generator polynomial $g(X) = 1 + X$ we have

$$
\begin{array}{r}
X^2 \qquad + 1 \\
\hline
X^3 + X^2 + X \\
X + 1 \, \bigg|\, X^3 + X^2 \\
\hline
\qquad X \\
\qquad X + 1 \\
\hline
\qquad 1
\end{array}
$$

The remainder $R(X) = 1$
Thus the code polynomial is

$$T(X) = X^3 + X^2 + X + 1$$

which is 1111

Proceeding as before, we get the following systematic code:

MESSAGES			CODE			
LSB		MSB	LSB			M3B
0	0	0	0	0	0	0
1	0	0	1	1	0	0
0	1	0	1	0	1	0
1	1	0	0	1	1	0
0	0	1	1	0	0	1
1	0	1	0	1	0	1
0	1	1	0	0	1	1
1	1	1	1	1	1	1

Assuming that message bits are input as

$M_1 M_2 0 M_1' M_2' M_3' M_0 \ldots,$

The code can be implemented by the following circuit.

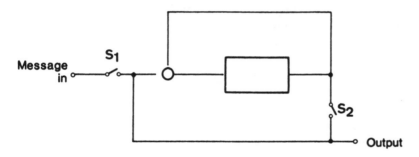

Figure A1.11

Initially, S2 is off and S1 is on. After M_3 is fed into the circuit, S1 switches off, S2 turns on and an additional shift by the SR generates the parity check code. Then, S2 switches off and S1 turns on to repeat the process.

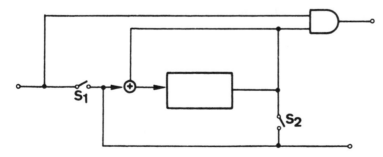

Figure A1.12

The decoder operates in the same manner as the encoder. However, an AND gate is inserted to check the parity received and the one generated by the SR. If they do not match, an error is detected.

Erasure and B-adjacent decoding

After the CRCC decoder detects a code error in each data block, error correction is implemented in the parity decoder. First the **syndrome** S_{P0} (corrector) is calculated as follows:

$$S_{P0} = L_0 + R_0 + L_1 + L_2 + P_0$$

where

$$P_0 = L_0 + R_0 + L_1 + R_1 + L_2 + R_2$$

If there is no error, $S_{P0} = 0$, because in modulo-2 summation, adding any two identical quantities is **zero**.

On the other hand, suppose a single error R_1' (i.e., all 16 bits are incorrect) in a single message block is made. In this case, $S_{P0} = 1$.

If we add this to R_1', correction can be made because:

If $R_{1X} = 0, R_{1,X} = 1$ where R_{1X} is any bit in R_1
    $R_{1X} = 1, R_{1,X} = 0$ where $R_{1,X}$ is any bit in R_1,

and, in either case, we have $R_1' + S_{P0} = R_1$. However, if two errors occur, this method of correction – called the **erase method** – fails to recover the original information. Since interleaving is employed in the PCM-F1, original bits in a block are **shuffled** to the other blocks by a 16H delay. Effectively, this method is capable of correcting errors occurring successively in 2048 bits.

In B-ADJACENT decoding, the syndromes S_{P0} and S_{Q0} are calculated as follows:

$$S_{P0} = L_{0r} + R_{0r} + L_{1r} + L_{2r} + R_{2r} + P_{0r} \ldots \tag{4}$$

$$S_{00} = T^6 L_{0r} + T^5 R_{0r} + R^4 L_{1r} + T^3 R_{1r} + T^2 L_{2r} + Q_{0r} \ldots \tag{5}$$

where

$$P_0 = L_0 + R_0 + L_1 + R_1 + L_2 + R_2{}^*$$
$$Q_0 = T^6 L_0 + T^5 R_0 + T^4 L_1 + T^3 R_1 + T^2 L_2 + TR_{2r}{}^*$$

$^*L_{0r}, \ldots, R_{2r}, P_{0r}, Q_{0r}$ are the information received.

Suppose the received information R_{1r} and R_{2r} contain errors, then:

$$R_1' = R_1 + E_{R1} \ldots \ldots \qquad (6)$$
$$R_2' = R_2 + F_{R2} \qquad (7)$$

where R_1 and R_2 are the original messages transmitted

E_{R1} and E_{R2} are the errors added.

Substituting R_1' and R_2' in place of R_{1r} and R_{2r} in (4) and (5), we can easily get,

$$S_{P0} = E_{R1} + E_{R2} \ldots \ldots \qquad (8)$$
$$S_{Q0} = T^3 E_{R1} + T E_{R2} \ldots \ldots \qquad (9)$$

adding T^{-1} to both sides of (9), then

$$T^{-1} S_{Q0} = T^2 E_{R1} + E_{R2} \ldots \ldots \qquad (10)$$

adding S_{P0} to both sides of (10),

$$
\begin{aligned}
S_{P0} + T^{-1} S_{Q0} &= (T^2 E_{R1} + E_{R2}) + S_{p0} \\
&= (T^2 E_{R1} + E_{R2}) + (E_{R1} + E_{R2}) \\
&= (T^2 + I) E_{R1} \ldots \ldots \qquad (11)
\end{aligned}
$$

rearrange 11:

$$E_{R1} = (T^2 + I) - 1 (S_{P0} + T^{-1} S_{Q0}) \ldots \ldots \qquad (12)$$

from 8:

$$E_{R2} = E_{R1} + S_{P0} \ldots \ldots \qquad (13)$$

Before we go on, let us summarize the situation as follows: we first calculate the syndromes S_{P0} and S_{Q0} according to (4) and (5). After some algebraic manipulations (if only 2 or less errors are made during the transmission of data), the error pattern can be calculated by (12) and (13). Once these are known, we can correct the error by (6) and (7) where:

$$R_1 = R'_1 + E_{R1}$$
$$R_2 = R'_2 + E_{R2}$$

Note that in modulo-2 summation, since $1 + 1 = 0$, $1 = -1$. By this method, 4096 successive bit errors can be corrected.

In what follows we shall demonstrate how the decoding is done by going through an example.

Unfortunately, the matrix T consists of 14×14 elements. Even for a simple example such as shown below, a lot of algebraic manipulations are involved.

Therefore, in order to avoid the many steps of purely mechanical computation from clouding the main issue, we shall show only the important steps or results. Anyway, what is important here is to get a 'feel' of how decoding is carried out.

Example. We have the following data:

LSB														MSB
L_0	0	0	0	0	0	0	0	0	0	0	0	0	0	0
R_0	1	0	0	0	0	0	0	0	0	0	0	0	0	0
L_1	0	1	0	0	0	0	0	0	0	0	0	0	0	0
R_1	0	0	1	0	0	0	0	0	0	0	0	0	0	0
L_2	0	0	0	1	0	0	0	0	0	0	0	0	0	0
R_2	0	0	0	0	1	0	0	0	0	0	0	0	0	0
P_0	1	1	1	1	1	0	0	0	0	0	0	0	0	0
Q_0	0	0	0	0	0	0	1	0	0	0	0	0	0	0

Suppose that, after the data is transmitted, the received data becomes:

L_0	0	0	0	0	0	0	0	0	0	0	0	0	0	0
R_0	1	0	0	0	0	0	0	0	0	0	0	0	0	0
L_1	0	1	0	0	0	0	0	0	0	0	0	0	0	0
R_1'	0	0	1	1	1	0	0	0	0	0	0	0	0	0
L_2	0	0	0	1	0	0	0	0	0	0	0	0	0	0
R_2'	0	0	0	0	0	0	0	0	0	1	1	1	0	0
P_0	1	1	1	1	1	0	0	0	0	0	0	0	0	0
Q_0	0	0	0	0	0	0	1	0	0	0	0	0	0	0

We can find E_{R1} and E_{R2} as follows:
First, we have to find S_{P0} and S_{Q0}.

Using (4) it is easy to see that

$$
S_{P0} = \begin{bmatrix} 0 \\ 0 \\ 0 \\ 1 \\ 0 \\ 0 \\ 0 \\ 0 \\ 0 \\ 1 \\ 1 \\ 1 \\ 0 \\ 0 \end{bmatrix}
$$

Similarly, from (5) and the column matrixes $TR_2, T^2 L_2, \ldots, T^6 L_0$ listed in the previous section, we find,

$$
S_{Q0} = \begin{bmatrix} 0 \\ 0 \\ 0 \\ 0 \\ 0 \\ 1 \\ 1 \\ 1 \\ 0 \\ 0 \\ 1 \\ 1 \\ 1 \\ 0 \end{bmatrix}
$$

After going through some algebraic manipulations, we find $T - 1$, $(T^2 + 1)$ and finally $(T^2 + 1)^{-1}$. We have,

$$S_{P0} + T^{-1}S_{Q0} = \begin{bmatrix} 0 \\ 0 \\ 0 \\ 1 \\ 1 \\ 1 \\ 1 \\ 0 \\ 0 \\ 0 \\ 0 \\ 0 \\ 0 \\ 0 \end{bmatrix}$$

Moreover, using (12), we obtain:

$$E_{R1} = \begin{bmatrix} 0 \\ 0 \\ 0 \\ 1 \\ 1 \\ 0 \\ 0 \\ 0 \\ 0 \\ 0 \\ 0 \\ 0 \\ 0 \\ 0 \end{bmatrix}$$

and finally from (13),

$$
E_{R2} = \begin{bmatrix} 0 \\ 0 \\ 0 \\ 0 \\ 1 \\ 0 \\ 0 \\ 0 \\ 0 \\ 1 \\ 1 \\ 1 \\ 0 \\ 0 \end{bmatrix}
$$

Now, you can verify that by taking the modulo-2 summation of $R'1$ and E_{R1}, R_2, and R_{R2}, the correct transmitted data are obtained.

Notes:

1 $(T^2 + 1)^{-1}$ can be constructed as follows:
   We have,

$$
(T^2 + I)L_0 = \begin{bmatrix} a_1 + a_{13} \\ a_2 + a_{14} \\ a_1 + a_3 \\ a_2 + a_4 \\ a_3 + a_5 \\ a_4 + a_6 \\ a_5 + a_7 \\ a_6 + a_8 \\ a_7 + a_9 + a_{13} \\ a_8 + a_{10} + a_{14} \\ a_9 + a_{11} \\ a_{10} + a_{12} \\ a_{11} + a_{13} \\ a_{12} + a_{14} \end{bmatrix} = M
$$

Because $(T^2 = 1)^1 M = L_0$, then for the first row of $(T^2 + 1)^{-1}$ we must have

$$\begin{bmatrix} 0 & 0 & 1 & 0 & 1 & 0 & 1 & 0 & 1 & 0 & 1 & 0 & 1 & 0 \end{bmatrix}$$

so that

$$\begin{bmatrix} 0 & 0 & 1 & 0 & 1 & 0 & 1 & 0 & 1 & 0 & 1 & 0 & 1 & 0 \end{bmatrix} M = a_1$$

2 We can find T^{-1} and $(T^2 + 1)$ in a manner similar to the above.
3 It can be verified easily that $(T^2 + 1)^{-1}(T^2 + 1) = 1$.
   When we multiply the first row and the first matrix by the first column of the second one, we have,

$$(0.1) + (0.0) + (1.1) + (0.0) + (1.0) + (0.0) + (1.0) +$$
$$(0.0) + (1.0) + (0.0) + (1.0) + (0.0) + (1.0) + (0.0) = 1$$

Thus, the first element of the matrix is 1. Similarly, we can verify that all the other elements, except the ones at the diagonal, are zero.

$$Q_0 = T^6 L_0 + T^5 R_0 + T^4 L_1 + T^3 R_1 + T^2 L_2 + TR_2$$

$$T^{-1} = \begin{bmatrix}
0 & 1 & 0 & 0 & 0 & 0 & 0 & 0 & 0 & 0 & 0 & 0 & 0 & 0 \\
0 & 0 & 1 & 0 & 0 & 0 & 0 & 0 & 0 & 0 & 0 & 0 & 0 & 0 \\
0 & 0 & 0 & 1 & 0 & 0 & 0 & 0 & 0 & 0 & 0 & 0 & 0 & 0 \\
0 & 0 & 0 & 0 & 1 & 0 & 0 & 0 & 0 & 0 & 0 & 0 & 0 & 0 \\
0 & 0 & 0 & 0 & 0 & 1 & 0 & 0 & 0 & 0 & 0 & 0 & 0 & 0 \\
0 & 0 & 0 & 0 & 0 & 0 & 1 & 0 & 0 & 0 & 0 & 0 & 0 & 0 \\
0 & 0 & 0 & 0 & 0 & 0 & 0 & 1 & 0 & 0 & 0 & 0 & 0 & 0 \\
0 & 0 & 0 & 0 & 0 & 0 & 0 & 0 & 1 & 0 & 0 & 0 & 0 & 0 \\
0 & 0 & 0 & 0 & 0 & 0 & 0 & 0 & 0 & 1 & 0 & 0 & 0 & 0 \\
0 & 0 & 0 & 0 & 0 & 0 & 0 & 0 & 0 & 0 & 1 & 0 & 0 & 0 \\
0 & 0 & 0 & 0 & 0 & 0 & 0 & 0 & 0 & 0 & 0 & 1 & 0 & 0 \\
0 & 0 & 0 & 0 & 0 & 0 & 0 & 0 & 0 & 0 & 0 & 0 & 1 & 0 \\
0 & 0 & 0 & 0 & 0 & 0 & 0 & 0 & 0 & 0 & 0 & 0 & 0 & 1 \\
1 & 0 & 0 & 0 & 0 & 0 & 0 & 0 & 0 & 0 & 0 & 0 & 0 & 0
\end{bmatrix}$$

$$T^2 + I = \begin{bmatrix}
1 & 0 & 0 & 0 & 0 & 0 & 0 & 0 & 0 & 0 & 0 & 0 & 1 & 0 \\
0 & 1 & 0 & 0 & 0 & 0 & 0 & 0 & 0 & 0 & 0 & 0 & 0 & 1 \\
1 & 0 & 1 & 0 & 0 & 0 & 0 & 0 & 0 & 0 & 0 & 0 & 0 & 0 \\
0 & 1 & 0 & 1 & 0 & 0 & 0 & 0 & 0 & 0 & 0 & 0 & 0 & 0 \\
0 & 0 & 1 & 0 & 1 & 0 & 0 & 0 & 0 & 0 & 0 & 0 & 0 & 0 \\
0 & 0 & 0 & 1 & 0 & 1 & 0 & 0 & 0 & 0 & 0 & 0 & 0 & 0 \\
0 & 0 & 0 & 0 & 1 & 0 & 1 & 0 & 0 & 0 & 0 & 0 & 0 & 0 \\
0 & 0 & 0 & 0 & 0 & 1 & 0 & 1 & 0 & 0 & 0 & 0 & 0 & 0 \\
0 & 0 & 0 & 0 & 0 & 0 & 1 & 0 & 1 & 0 & 0 & 0 & 0 & 0 \\
0 & 0 & 0 & 0 & 0 & 0 & 0 & 1 & 0 & 1 & 0 & 0 & 0 & 0 \\
0 & 0 & 0 & 0 & 0 & 0 & 0 & 0 & 1 & 0 & 1 & 0 & 0 & 0 \\
0 & 0 & 0 & 0 & 0 & 0 & 0 & 0 & 0 & 1 & 0 & 1 & 0 & 0 \\
0 & 0 & 0 & 0 & 0 & 0 & 0 & 0 & 0 & 0 & 1 & 0 & 1 & 0 \\
0 & 0 & 0 & 0 & 0 & 0 & 0 & 0 & 0 & 0 & 0 & 1 & 0 & 1
\end{bmatrix}$$

$$(T^2 + I)^{-1} = \begin{bmatrix}
0 & 0 & 1 & 0 & 1 & 0 & 1 & 0 & 1 & 0 & 1 & 0 & 1 & 0 \\
0 & 0 & 0 & 1 & 0 & 1 & 0 & 1 & 0 & 1 & 0 & 1 & 0 & 1 \\
0 & 0 & 0 & 0 & 1 & 0 & 1 & 0 & 1 & 0 & 1 & 0 & 1 & 0 \\
0 & 0 & 0 & 0 & 0 & 1 & 0 & 1 & 0 & 1 & 0 & 1 & 0 & 1 \\
0 & 0 & 0 & 0 & 0 & 0 & 1 & 0 & 1 & 0 & 1 & 0 & 1 & 0 \\
0 & 0 & 0 & 0 & 0 & 0 & 0 & 1 & 0 & 1 & 0 & 1 & 0 & 1 \\
0 & 0 & 0 & 0 & 0 & 0 & 0 & 0 & 1 & 0 & 1 & 0 & 1 & 0 \\
0 & 0 & 0 & 0 & 0 & 0 & 0 & 0 & 0 & 1 & 0 & 1 & 0 & 1 \\
1 & 0 & 1 & 0 & 1 & 0 & 1 & 0 & 1 & 0 & 0 & 0 & 0 & 0 \\
0 & 1 & 0 & 1 & 0 & 1 & 0 & 1 & 0 & 1 & 0 & 0 & 0 & 0 \\
1 & 0 & 1 & 0 & 1 & 0 & 1 & 0 & 1 & 0 & 1 & 0 & 0 & 0 \\
0 & 1 & 0 & 1 & 0 & 1 & 0 & 1 & 0 & 1 & 0 & 1 & 0 & 0 \\
1 & 0 & 1 & 0 & 1 & 0 & 1 & 0 & 1 & 0 & 1 & 0 & 1 & 0 \\
0 & 1 & 0 & 1 & 0 & 1 & 0 & 1 & 0 & 1 & 0 & 1 & 0 & 1
\end{bmatrix}$$

Appendix 2

Anatomy of a PAL Composite Video Signal (Monochrome)

Within a period of 64 μs, the digital data processing circuit of the PCM-F1 outputs 168 bits of data. If these data are to be recorded on tape, the tape recorder has to handle about 2.6×10^6 bit s^{-1}. Presently, only video tape recorders have the required bandwidth to perform this task, which means that digital data have to be turned into a video signal before it can be recorded.

This appendix looks at the components of a composite video signal. At the end, a list of video signal parameters and definitions of video terms are included as reference.

In a television set, two independent free-running oscillators are employed to generate sawtooth voltages that deflect an electron beam horizontally and vertically over the entire display screen. The sweep is from left to right, and top to bottom. The time required for an electron beam to make a horizontal 'round-trip', i.e., from the left edge to the right edge and then back is defined as H. Similarly, a vertical round-trip is termed V.

The frequency of H intervals per second, called the horizontal frequency, is set at 15,625 Hz. To minimize interference on the television screen caused by beating, the vertical frequency (i.e., intervals per second) is chosen to be the same as the frequency of mains power: 50 Hz. (Although modern TVs are immune to mains power beat interference, the 50 Hz standard remains.) These two frequencies determine the line and field rates of the display. From these two numbers, we get, in 1 field, 15,625/50 = 312.5 lines. Also, 1H occurs in a period of 64 μs.

For meaningful pictures to be displayed on the screen, the free-running oscillators must be triggered by some means so that they are synchronized to

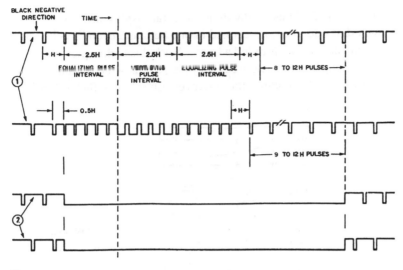

① SYNC SIGNAL

② BLANKING SIGNAL

Figure A2.1

both line and field rates. Hence, both horizontal and vertical sync pulses are inserted into the video signal. Refer to figure A2.2. However, the television receiver must have some way of detecting which sync pulse is being received at any one time. Figure A2.2a shows horizontal sync pulses at the end of one field and the beginning of the next. Between the times T_1 and T_2, therefore, vertical sync pulses must be inserted, and must be sufficiently different from the horizontal sync pulses so that the receiver knows the difference. The technique is to use pulses of a different width, called field broad pulses (Figure A2.2b). Thus a simple differentiator circuit will detect horizontal sync pulses, while a similarly simple integrator circuit will detect vertical sync pulses.

As we have mentioned before, in one field, 312.5 lines are scanned. However, the horizontal resolution of the raster can be increased,

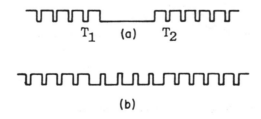

T_1 (a) T_2

(b)

Figure A2.2 *Sync signals in a PAL signal (a) horizontal sync signals at the end of one field and the beginning of the next (b) composite sync signal with field broad pulses (vertical sync pulses) added*

theoretically, to 625 lines by dividing the total display into two interlaced fields, such that odd lines are scanned in one field and even lines in another. The two fields form what is called a frame. With this technique, although the frame rate is only half the field frequency i.e., 25Hz, which could give rise to annoying flicker of the picture, the raster refresh rate is a 'flicker free' 50Hz.

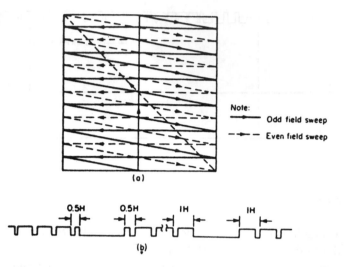

Figure A2.3 *Interlaced fields (a) interlaced display (b) using a ½H shift in the vertical sync pulses*

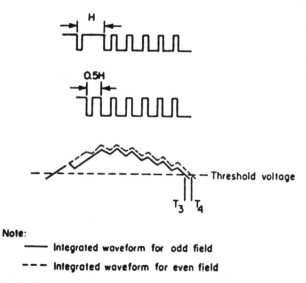

Figure A2.4 *Problems created in timing when a ½H shift in vertical sync pulses is used to interlace fields*

Once interlacing is adopted, the composite sync signal gets more complicated. The sweep and retrace pattern, shown in Figure A.2.3a, can be generated by the sync signal indicated in Figure A2.3b. Note that it is the ½H shift in the vertical sync pulses which interlaces the fields.

The ½H shift in vertical sync pulses, however, creates a timing problem in that each field (odd and even) of the frame will be triggered at different times (T_3 and T_4). This is illustrated in Figure A2.4. The solution is to add extra pulses before and after vertical sync pulses to ensure there is a similar pulse pattern around the vertical sync pulses for each field (Figure A2.5).

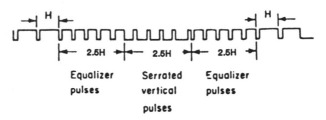

Figure A2.5 *Addition of equalizing pulses to solve the field timing problem*

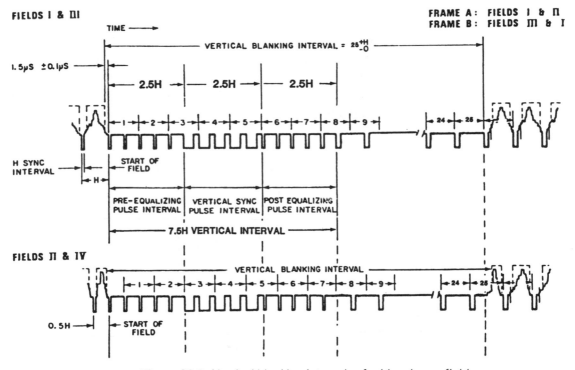

Figure A2.6 *Vertical blanking intervals of odd and even fields*

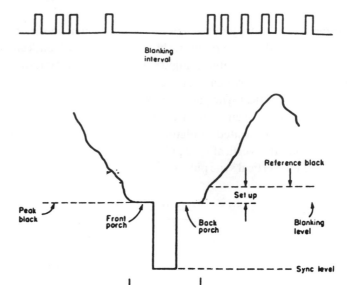

Figure A2.7 *Horizontal line sync signals are combined with analog display data during the display time*

All the pulses discussed are combined in Figure A2.6, where the vertical blanking intervals of odd and even fields of a composite video signal are shown. Figure A2.7 shows how horizontal live sync pulse and analog display data are combined for the rest of the signal, along with nomenclature. Finally, Table A2.1 lists main specifications of the PAL and NSTC composite video signal formats.

Table A2.1 *Main specification of PAL and NTSC composite video signals*

	PAL	NTSC
Lines per frame	625	525
Field rate (Hz)	50	60
Line period (μs)	64	63.5
Field period (ms)	20	16.67
Horizontal sync pulse (μs)	5.8	5.1
Vertical sync pulse (μs)	160	158.75
Horizontal blanking (μs)	12.05	11.4
Vertical blanking (ms)	1.28	1.27

Definition of video terms

Blanking level
The level which shuts off CRT current, resulting in the blackest possible picture.

Composite sync signal
The portion of the composite video signal which synchronizes the scanning process.

Composite video signal
The display signal plus the composite sync signal.

Reference black level
The maximum peak excursion of the display signal in the negative direction.

Reference white level
The maximum peak excursion of the display signal in the positive direction.

Setup
The difference between the reference black level and the blanking level.

Sync level
The peak level of the composite sync signal.

Appendix 3

Sampling theorem

Sampling picks out values $f(nT)$ from a signal $f(t)$, at regular intervals. This is equivalent to the multiplication of $f(t)$ with the signal $s(t)$, given by the expression:

$$fs(t) = \sum_{n=-\infty}^{\infty} Tf(nT)\delta(t-nT)$$

where (t) is a delta function.

The Fourier transform $F_s(\omega)$ of $F_s(t)$, is given by the expression:

$$F_s(\omega) = \int_{-\infty}^{\infty} \sum_{n=-\infty}^{\infty} Tf(nT)\delta(t-nT)e^{-j\omega t}\,dt$$

where $\omega_s = 2\pi/T$. When two functions are multiplied in time their transforms in the frequency domain are convoluted. For this reason, the spectrum $F(\omega)$ is repeated at multiples of the sampling frequency. Function $f(t)$ may be recovered from $F_s(\omega)$ by first multiplying by a gating function $G(\omega)$, illustrated in Figure A3.1. This results in the expression:

$$F(\omega) = F_s(\omega)G(\omega)$$

Now, the inverse transform of $G(\omega)$ is given by the expression:

$$\frac{\sin \omega_0 t}{t}$$

and if:

$$F_s(\omega) \leftrightarrow f_s(t)$$

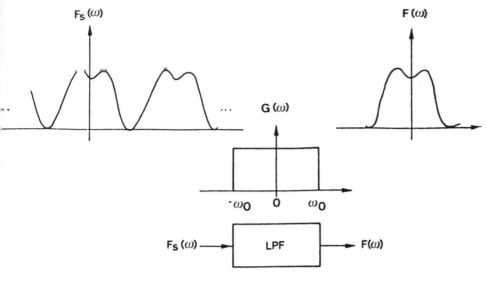

Figure A3.1 *Recovery of a sampled signal*

then:

$$G(\omega) \leftrightarrow g(t)$$

so that:

$$F_s(\omega) G(\omega) \leftrightarrow f_s(t) * g(t)$$

where the symbol * denotes convolution. This gives:

$$f(t) = \sum_{n=-\infty}^{\infty} Tf(nT) \delta(t - nT) * \frac{\sin \omega_0 t}{T_0 \omega_0 t/2}$$

$$= \sum_{n=-\infty}^{\infty} Tf(nT) \frac{\sin \omega_0(t - nT)}{T_0 \omega_0(t - nT)/2}$$

and, when: $T_0 = 2T$:

$$f(t) = \sum_{n=-\infty}^{\infty} f(nT) \frac{\sin \omega_0(t - nT)}{\omega_0(t - nT)}$$

This result, the **sampling theorem**, relates the samples $f(nT)$ taken at regular intervals to the function $f(t)$.

The sampling theorem is important when considering the bandwidth of a sampled signal. For example, a function f(t) can only be properly reconstructed when samples have been taken at the **Nyquist rate**, $1/T = 2/T_0$. In practical terms, this means that the sampling frequency should be twice that of the highest signal component frequency, i.e.:

$$F_s = 2f_0$$

and so, to make sure that signal component frequencies greater than half the sampling frequency are not sampled, an anti-aliasing filter is used.

Appendix 4

Serial Copy Management System (SCMS)

The introduction of new digital recording systems like DAT, MiniDisc and DCC made it perfectly possible to produce high quality copies of existing materials. When using digital interconnection, there is virtually no sound quality degradation. This raises the question for protection of copyrighted material. The SCMS limits the ability of consumer digital audio recording equipment to make digital transfers. SCMS allows a single generation copy of a copyrighted digital source. For example, a copyrighted CD can be digitally copied to a DAT recorder once (and only once). The SCMS system prevents further digital copies of that tape. Figure A4.1 shows some examples of possible situations. Note that regardless of the source, it is always possible to make copies by using analog connections.

Three possible cases exist when digitally copying a digital source.

- The source is not copyrighted: digital copy permitted
- The source is copyrighted: a first generation copy is permitted
- The source is a digital copy of copyrighted material: digital copy prohibited.

Each of the digital recording systems contain some auxiliary data, needed for correct recovery of the digital audio signal. The SCMS copy bits can be found in the auxiliary data, e.g. in case of DAT they are contained in ID6 bit 6 and bit 7 (see Table 17.4). In the Digital Audio Interface Format (S/PDIF) the copy bits are included in the channel status bits.

The SCMS copy management routine is shown in Fig. A4.2. The SCMS software checks the channel status bit of the digital audio signal. Several requirements must be fulfilled on the recording side before copying can

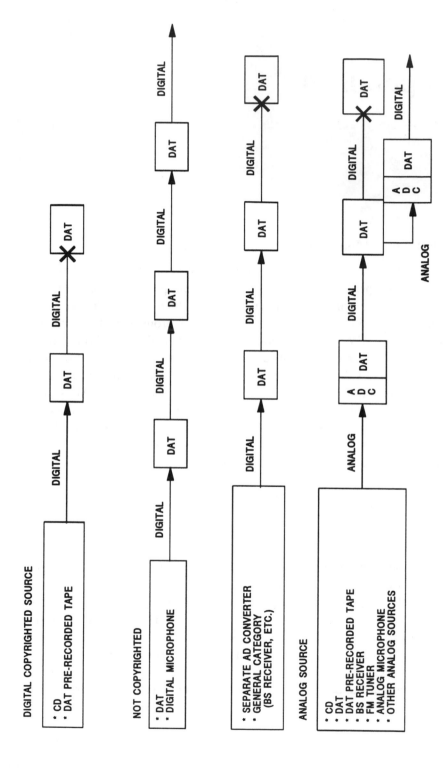

Figure A4.1 *Different situations when copying via the SCMS system*

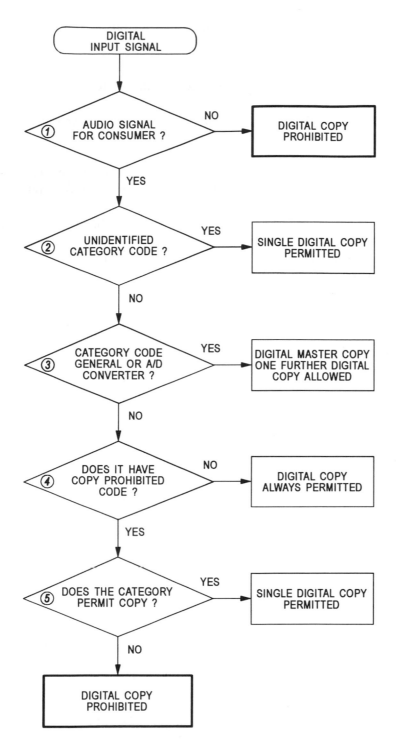

Figure A4.2 *SCMS management routine*

start. The recording machine will alter the copy bits according to the new situation.

Note the following.

● A digital copy is never possible when the digital audio signal is not for consumer use
● When the category code is unidentified, the source is considered copyrighted. A single digital copy is allowed, but new copy bits will be set to prevent further copies
● When the category code is 'general' or 'A/D converter', the recording is considered to be a (copyrighted) master recording. The copy bits will be set to allow one further digital copy
● If the channel status bits do not carry a copy prohibited code, the source is considered not copyrighted. Digital copies will always be possible
● Even when the copy prohibited code is set, certain categories (e.g. CD) allow for a single digital copy with the new copy bits set to copy prohibited
● In all other cases a digital copy is prohibited.

Appendix 5

Digital Audio Interface Format (S/PDIF–IEC958)

An interface format is designed for use between digital audio sets. This format was originally defined jointly by Sony and Philips (Sony Philips Digital Interface Format—S/PDIF) but has become an international standard.

This format allows digital connection of digital audio sets, such as CD players, DAT players and others, thus excluding any quality loss related to multiple D/A–A/D conversion.

Transmission method can be either standard (coax cable) or optical.

In the S/PDIF format, both left and right channels are transmitted on one line, they are multiplexed.

Note that the SCMS system is included in this S/PDIF format in order to avoid illegal copying. Figure A5.1 shows the S/PDIF data format.

Each subframe will give full information on one audio sample. If the audio sample is less than 20 bits, the remaining bits are meaningless. Two sub-frames make one frame, giving a left and right channel sample.

192 frames make one block, the channel status bits of one block, i.e. 192 bits taken from each channel bit in each subframe give more information about type of audio, sampling rate etc.

Channel Coding

For transmission of the S/PDIF format, a channel coding/biphase modulation type was chosen to ensure proper clocking. If the input data is '1', there is a transient in the middle of the bit; if the input data is '0', there is no transient in the middle of the bit; at the end of each bit there will also be a transient. Figure A5.2 shows the channel coding.

The preambles of each subframe are encoded in a unique way, as shown in Fig. A5.3.

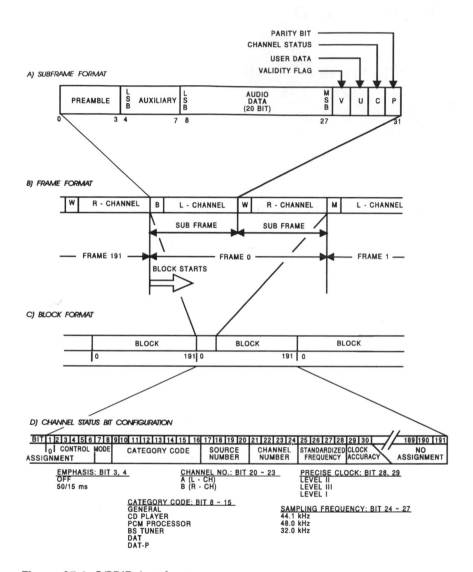

Figure A5.1 *S/PDIF data format*

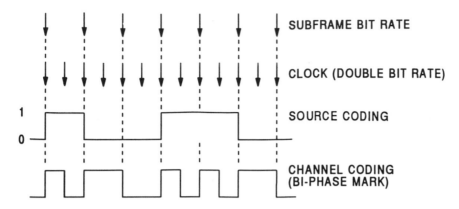

Figure A5.2 *Channel coding*

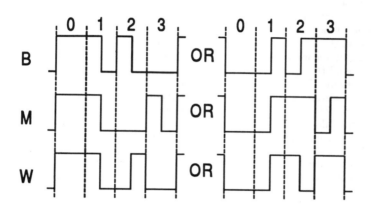

Figure A5.3 *Preamble coding*

Index